Lecture Notes in Mathematics

For information about Vols. 1–925, please contact your bookseller or Springer-Verlag.

Vol. 926: Geometric Techniques in Gauge Theories. Proceedings, 1981. Edited by R. Martini and E.M.de Jager. IX, 219 pages. 1982.

Vol. 927: Y. Z. Flicker, The Trace Formula and Base Change for GL (3). XII, 204 pages. 1982.

Vol. 928: Probability Measures on Groups. Proceedings 1981. Edited by H. Heyer. X, 477 pages. 1982.

Vol. 929: Ecole d'Eté de Probabilités de Saint-Flour X – 1980. Proceedings, 1980. Edited by P.L. Hennequin. X, 313 pages. 1982.

Vol. 930: P. Berthelot, L. Breen, et W. Messing, Théorie de Dieudonné Cristalline II. XI, 261 pages. 1982.

Vol. 931: D.M. Arnold, Finite Rank Torsion Free Abelian Groups and Rings. VII, 191 pages. 1982.

Vol. 932: Analytic Theory of Continued Fractions. Proceedings, 1981. Edited by W.B. Jones, W.J. Thron, and H. Waadeland. VI, 240 pages. 1982.

Vol. 933: Lie Algebras and Related Topics. Proceedings, 1981. Edited by D. Winter. VI, 236 pages. 1982.

Vol. 934: M. Sakai, Quadrature Domains. IV, 133 pages. 1982.

Vol. 935: R. Sot, Simple Morphisms in Algebraic Geometry. IV, 146 pages. 1982.

Vol. 936: S.M. Khaleelulla, Counterexamples in Topological Vector Spaces. XXI, 179 pages. 1982.

Vol. 937: E. Combet, Intégrales Exponentielles. VIII, 114 pages. 1982.

Vol. 938: Number Theory. Proceedings, 1981. Edited by K. Alladi. IX, 177 pages. 1982.

Vol. 939: Martingale Theory in Harmonic Analysis and Banach Spaces. Proceedings, 1981. Edited by J.-A. Chao and W.A. Woyczyński. VIII, 225 pages. 1982.

Vol. 940: S. Shelah, Proper Forcing. XXIX, 496 pages. 1982.

Vol. 941: A. Legrand, Homotopie des Espaces de Sections. VII, 132 pages. 1982.

Vol. 942: Theory and Applications of Singular Perturbations. Proceedings, 1981. Edited by W. Eckhaus and E.M. de Jager. V, 363 pages. 1982.

Vol. 943: V. Ancona, G. Tomassini, Modifications Analytiques. IV, 120 pages. 1982.

Vol. 944: Representations of Algebras. Workshop Proceedings, 1980. Edited by M. Auslander and E. Lluis. V, 258 pages. 1982.

Vol. 945: Measure Theory. Oberwolfach 1981, Proceedings. Edited by D. Kölzow and D. Maharam-Stone. XV, 431 pages. 1982.

Vol. 946: N. Spaltenstein, Classes Unipotentes et Sous-groupes de Borel. IX, 259 pages. 1982.

Vol. 947: Algebraic Threefolds. Proceedings, 1981. Edited by A. Conte. VII, 315 pages. 1982.

Vol. 948: Functional Analysis. Proceedings, 1981. Edited by D. Butković, H. Kraljević, and S. Kurepa. X, 239 pages. 1982.

Vol. 949: Harmonic Maps. Proceedings, 1980. Edited by R.J. Knill, M. Kalka and H.C.J. Sealey. V, 158 pages. 1982.

Vol. 950: Complex Analysis. Proceedings, 1980. Edited by J. Eells. IV, 428 pages. 1982.

Vol. 951: Advances in Non-Commutative Ring Theory. Proceedings, 1981. Edited by P.J. Fleury. V, 142 pages. 1982.

Vol. 952: Combinatorial Mathematics IX. Proceedings, 1981. Edited by E. Billington, S. Oates-Williams, and A.P. Street. XI, 443 pages. 1982.

Vol. 953: Iterative Solution of Nonlinear Systems of Equations. Proceedings, 1982. Edited by R. Ansorge, Th. Meis, and W. Törnig. VII, 202 pages. 1982.

Vol. 954: S.G. Pandit, S.G. Deo, Differential Systems Involving Impulses. VII, 102 pages. 1982.

Vol. 955: G. Gierz, Bundles of Topological Vector Spaces and Their Duality. IV, 296 pages. 1982.

Vol. 956: Group Actions and Vector Fields. Proceedings, 1981. Edited by J.B. Carrell. V, 144 pages. 1982.

Vol. 957: Differential Equations. Proceedings, 1981. Edited by D.G. de Figueiredo. VIII, 301 pages. 1982.

Vol. 958: F.R. Beyl, J. Tappe, Group Extensions, Representations, and the Schur Multiplicator. IV, 278 pages. 1982.

Vol. 959: Géométrie Algébrique Réelle et Formes Quadratiques, Proceedings, 1981. Edité par J.-L. Colliot-Thélène, M. Coste, L. Mahé, et M.-F. Roy. X, 458 pages. 1982.

Vol. 960: Multigrid Methods. Proceedings, 1981. Edited by W. Hackbusch and U. Trottenberg. VII, 652 pages. 1982.

Vol. 961: Algebraic Geometry. Proceedings, 1981. Edited by J.M. Aroca, R. Buchweitz, M. Giusti, and M. Merle. X, 500 pages. 1982.

Vol. 962: Category Theory. Proceedings, 1981. Edited by K.H. Kamps D. Pumplün, and W. Tholen, XV, 322 pages. 1982.

Vol. 963: R. Nottrot, Optimal Processes on Manifolds. VI, 124 pages. 1982.

Vol. 964: Ordinary and Partial Differential Equations. Proceedings, 1982. Edited by W.N. Everitt and B.D. Sleeman. XVIII, 726 pages. 1982.

Vol. 965: Topics in Numerical Analysis. Proceedings, 1981. Edited by P.R. Turner. IX, 202 pages. 1982.

Vol. 966: Algebraic K-Theory. Proceedings, 1980, Part I. Edited by R.K. Dennis. VIII, 407 pages. 1982.

Vol. 967: Algebraic K-Theory. Proceedings, 1980. Part II. VIII, 409 pages. 1982.

Vol. 968: Numerical Integration of Differential Equations and Large Linear Systems. Proceedings, 1980. Edited by J. Hinze. VI, 412 pages. 1982.

Vol. 969: Combinatorial Theory. Proceedings, 1982. Edited by D. Jungnickel and K. Vedder. V, 326 pages. 1982.

Vol. 970: Twistor Geometry and Non-Linear Systems. Proceedings, 1980. Edited by H.-D. Doebner and T.D. Palev. V, 216 pages. 1982.

Vol. 971: Kleinian Groups and Related Topics. Proceedings, 1981. Edited by D.M. Gallo and R.M. Porter. V, 117 pages. 1983.

Vol. 972: Nonlinear Filtering and Stochastic Control. Proceedings, 1981. Edited by S.K. Mitter and A. Moro. VIII, 297 pages. 1983.

Vol. 973: Matrix Pencils. Proceedings, 1982. Edited by B. Kågström and A. Ruhe. XI, 293 pages. 1983.

Vol. 974: A. Draux, Polynômes Orthogonaux Formels – Applications. VI, 625 pages. 1983.

Vol. 975: Radical Banach Algebras and Automatic Continuity. Proceedings, 1981. Edited by J.M. Bachar, W.G. Bade, P.C. Curtis Jr., H.G. Dales and M.P. Thomas. VIII, 470 pages. 1983.

Vol. 976: X. Fernique, P.W. Millar, D.W. Stroock, M. Weber, Ecole d'Eté de Probabilités de Saint-Flour XI – 1981. Edited by P.L. Hennequin. XI, 465 pages. 1983.

Vol. 977: T. Parthasarathy, On Global Univalence Theorems. VIII, 106 pages. 1983.

Vol. 978: J. Ławrynowicz, J. Krzyż, Quasiconformal Mappings in the Plane. VI, 177 pages. 1983.

Vol. 979: Mathematical Theories of Optimization. Proceedings, 1981. Edited by J.P. Cecconi and T. Zolezzi. V, 268 pages. 1983.

continued on page 181

Lecture Notes in Mathematics

Edited by A. Dold and B. Eckmann

1129

Numerical Analysis Lancaster 1984

Proceedings of the SERC Summer School
held in Lancaster, England, Jul. 15 – Aug. 3, 1984

Edited by P.R. Turner

Springer-Verlag
Berlin Heidelberg New York Tokyo

Editor

Peter R. Turner
Department of Mathematics, University of Lancaster
Bailrigg, Lancaster LA1 4YL, England

Mathematics Subject Classification (1980): 35J25, 41A50, 41A63, 41A65, 65N15, 65N30

ISBN 3-540-15234-2 Springer-Verlag Berlin Heidelberg New York Tokyo
ISBN 0-387-15234-2 Springer-Verlag New York Heidelberg Berlin Tokyo

Printing and binding: Beltz Offsetdruck, Hemsbach / Bergstr.
2146/3140-543210

PREFACE

THE S.E.R.C. NUMERICAL ANALYSIS SUMMER SCHOOL

University of Lancaster
15th July - 3rd August, 1984

The aims of this second Numerical Analysis Summer School were much the same as for the inaugural event in 1981 - LNM 965 describes the proceedings of the first one - although the organizational structure was altered. Each week of this conference was essentially self-contained. During each week there was provision for intensive study, which could broaden participants' research interests or deepen their understanding of topics with which they already had some familiarity, and for continuing individual research in the stimulating environment created by the gathering together of several experts of international renown.

This volume contains lecture notes for most of the main courses of lectures given at the meeting. Each week there was one ten-lecture course, a related course of five lectures and a programme of research seminars given by further invited speakers or other participants. The one week for which lecture notes are not included is that on Multigrid Methods. The courses in that week were based on material which has appeared in W. Hackbusch and U. Trottenberg (eds) Multigrid Methods, LNM 960, Springer-Verlag, Heidelberg, 1984. Detailed lists of contents of the papers of Hackbusch and Brandt are included here. Brandt's extensive "Multigrid Guide" is updated annually and is available as a GMD monograph.

Presented here then is an account of the proceedings of a very successful and well-attended second meeting in what it is hoped will become a regular series of Numerical Analysis Summer Schools at Lancaster.

Acknowledgements

The first and most important acknowledgement is to the Science and Engineering Research Council who sponsored the meeting very generously. Their contribution covered all the organizational and running costs, the full expenses of the invited experts and the accommodation and subsistence expenses for twenty participants each week. A contribution to the expenses of the meeting was also received from Marconi Underwater Systems Ltd. I also wish to thank the Mathematics Secretariat of SERC and Professor John Whiteman, who acted as their assessor, as well as Professor Charles Clenshaw and other colleagues at Lancaster for their help and encouragement.

Finally, a word of thanks is due to Marion Garner who handled virtually all the secretarial work during the preparation and running of the meeting. Her help was invaluable.

Peter R. Turner,
Department of Mathematics,
University of Lancaster.

MAIN SPEAKERS

Prof. A. Brandt, Department of Applied Mathematics, The Weizmann Institute of Science, Rehovot, Israel.

Prof. E.W. Cheney, Department of Mathematics, University of Texas at Austin, Austin, Texas, U.S.A.

Prof. W. Hackbusch, Informatik, Christian Albrechts Universitaat, Kiel, West Germany.

Dr. C.A. Micchelli, IBM Thomas J. Watson Research Center, Box 218, Yorktown Heights, N.Y., U.S.A.

Dr. T.J. Rivlin, IBM Thomas J. Watson Research Center, Box 218, Yorktown Heights, N.Y. U.S.A.

Prof. A. Schatz, Department of Mathematics, Cornell University, Ithaca, N.Y., U.S.A.

Prof. R.E. Showalter, Department of Mathematics, The University of Texas at Austin, Austin, Texas, U.S.A.

INVITED SPEAKERS

Dr. C.T.H. Baker, Department of Mathematics, The University, Manchester.

Prof. C.W. Clenshaw, Department of Mathematics, University of Lancaster, Lancaster.

Dr. D. Kershaw, Department of Mathematics, University of Lancaster, Lancaster.

Prof. K.W. Morton, Oxford University Computing Laboratory, Oxford.

Prof. M.J.D. Powell, DAMTP, The University, Silver Street, Cambridge.

Prof. J.R. Whiteman, Institute of Computational Mathematics, Brunel University, Uxbridge, Middlesex.

PARTICIPANTS

Prof. M.A. Akhlagi, Mathematics Department, Shiraz University, Shiraz, Iran.

Dr. S. Amini, Department of Mathematics, Statistics and Computing, Plymouth Polytechnic, Plymouth.

Mr. P.J. Aston, Department of Mathematics and Statistics, Brunel University, Uxbridge, Middlesex.

Dr. M.J. Baines, Department of Mathematics, University of Reading, Reading.

Dr. J.W. Barrett, Department of Mathematics, Imperial College, London.

Mr. A. Beagles, Department of Mathematics and Statistics, Brunel University, Uxbridge, Middlesex.

Mr. R.P. Bennell, Mathematics Branch, R.M.C.S., Shrivenham, Swindon, Wilts.

Dr. M. Berzins, Department of Computer Studies, The University, Leeds.

Mr. S. Bishop, Mathematics Department, Polytechnic of North London, London.

Dr. J.S. Bramley, Department of Mathematics, University of Strathclyde, Glasgow.

Dr. A.L. Brown, School of Mathematics, The University, Newcastle upon Tyne.

Mr. P. Campbell, Marconi Underwater Systems Ltd., Wembley, Middlesex.

Dr. K.W.E. Chu, Department of Mathematics, University of Reading, Reading.

Mr. S. Chynoweth, Meteorological Office, Bracknell, Berks.

Dr. A.W. Craig, Department of Civil Engineering, University College of Swansea, Swansea.

Mr. M.P. Cullinan, DAMTP, The University, Cambridge.

Miss A. Daman, Mathematics Branch, R.M.C.S., Shrivenham, Swindon.

Mr. I.C. Demetriou, DAMTP, The University, Cambridge.

Mr. R.P. Dixon, Department of Mathematics, University of Reading, Reading.

Mr. P. Dolan, The Numerical Optimisation Centre, Hatfield Polytechnic, Hatfied, Herts.

Mr. T. Edgecock, Marconi Underwater Systems Ltd., Wembley, Middlesex.

Mr. N. Edwards, C.E.R.L., Leatherhead.

Dr. C.M. Elliott, Department of Mathematics, Imperial College, London.

Mr. S. Ellis, Department of Mathematics, Coventry Polytechnic, Coventry.

Dr. A. Ghafarian, Department of Mathematics, University of Kerman, Kerman, Iran.

Mr. J. Gilbert, Department of Mathematics, University of Lancaster, Lancaster.

Prof. M.v. Golitschek, Institut fur Angewandte Mathematik, Wurzburg, West Germany.

Dr. T.N.T. Goodman, Department of Mathematical Sciences, The University, Dundee.

Mr. E.W. Haddon, School of Computing Studies and Accountancy, University of East Anglia, Norwich.

Dr. D.C. Handscomb, Oxford University Computing Laboratory, Oxford.

Mr. H. Hanse, Department of Mathematics, Institute of Technology, Lulea, Sweden.

Dr. P.J. Harley, Department of Applied and Computational Mathematics, University of Sheffield, Sheffield.

Prof. W. Haussmann, Department of Mathematics, University of Duisburg, Duisburg, West Germany.

Dr. D.M. Hough, Department of Mathematics, Coventry Polytechnic, Coventry.

Mr. R.A. Hughes, Department of Mathematics, University of Bristol, Bristol.

Mr. P. Jackson, Mathematics Branch, R.M.C.S., Shrivenham, Swindon, Wilts.

Dr. N. Jacob, Hochschule der Bundeswehr Munchen, Institut fur Mathematik, Neubiberg, W. Germany.

Dr. A. King, Marconi Underwater Systems Ltd., Wembley, Middlesex.

Dr. W.A. Light, Department of Mathematics, University of Lancaster, Lancaster.

Mr. R. Löhner, Department of Civil Engineering, University College of Swansea, Swansea.

Dr. J.C. Mason, Mathematics Branch, R.M.C.S., Shrivenham, Swindon, Wilts.

Dr. G.P. McKeown, School of Computing Studies and Accountancy, University of East Anglia, Norwich.

Dr. J.L. Mohamed, Department of Statistics and Computational Mathematics, The University, Liverpool.

Dr. G. Moore, Department of Mathematics and Statistics, Brunel University, Uxbridge, Middlesex.

Dr. J.A. Murphy, Department of Mathematics, University of Aston, Birmingham.

Mr. M. Paisley, Department of Mathematics, University of Reading, Reading.

Mr. V. Pau, Department of Mathematics, University of Bristol, Bristol.

Dr. W.M. Pickering, Department of Applied and Computational Mathematics, University of Sheffield, Sheffield.

Dr. S.C. Power, Department of Mathematics, University of Lancaster, Lancaster.

Mr. R. Preston, C.E.R.L., Leatherhead.

Dr. J.D. Pryce, Department of Computer Science, University of Bristol, Bristol.

Mrs. S. Rose, Mathematics Branch, R.M.C.S., Shrivenham, Swindon, Wilts.

Mrs. J. Scott, Oxford University Computing Laboratory, Oxford.

Prof. A. Sharma, Mathematics Department, University of Alberta, Edmonton, Canada.

Mr. G. Shaw, Oxford University Computing Laboratory, Oxford.

Mr. J.E.H. Shaw, Department of Mathematics, University of Nottingham, Nottingham.

Mr. E.E. Sili, Department of Mathematics, University of Reading, Reading.

Miss C. Sims, Department of Mathematics, University of Lancaster, Lancaster.

Dr. P. Singh, Oxford University Computing Laboratory, Oxford.

Mr. R. Stenberg, Department of Mathematics, Helsinki University of Technology, Espoo, Finland.

Miss E.J. Stern, Mathematics Branch, R.M.C.S., Shrivenham, Swindon, Wilts.

Dr. R.M. Thomas, Department of Mathematics, UMIST, Manchester.

Miss S.A. Trickett, Oxford University Computing Laboratory, Oxford.

Dr. P.R. Turner, Department of Mathematics, University of Lancaster, Lancaster.

Dr. E.H. Twizell, Department of Mathematics and Statistics, Brunel University, Uxbridge, Middlesex.

Miss J.M. Wilkins, Mathematics Institute, University of Kent at Canterbury, Canterbury.

Dr. K. Wright, University Computing Laboratory, University of Newcastle, Newcastle-upon-Tyne.

LECTURE PROGRAMME

(a) Ten-lecture courses.

A. Brandt, Multigrid Techniques.

C.A. Micchelli and T.J. Rivlin, Optimal methods in approximation theory.

R.E. Showalter, Variational theory and approximation of boundary value problems.

(b) Five-lecture courses.

E.W. Cheney, Algorithmic aspects of approximation theory.

W. Hackbusch, Convergence analysis of multigrid methods.

A. Schatz, An introduction to the analysis of the error in the finite element method for second-order elliptic boundary value problems.

(c) Seminar programme.

M.J. Baines, 1. Algorithms for advection and shock problems in one and two dimensions.
2. Moving finite elements for scalar hyperbolic equations.

C.T.H. Baker, Approximations to the exponential function arising in the numerical analysis of Volterra integral equations.

J.W. Barrett and C.M. Elliott, Variational crimes in the finite element method (2).

A.L. Brown, Best approximations by smooth functions and related problems.

C.W. Clenshaw, Manipulations of large numbers (2).

A.W. Craig, Hierarchical finite element bases and multigrid methods.

M.P. Cullinan, Smoothing of data by removing sign changes in divided differences of a prescribed order.

M.v. Golitschek, Shortest path algorithms in bivariate approximation theory.

T.N.T. Goodman, The β-spline : a tool in computer aided design.

D.C. Handscomb, Recovery of smooth solutions from finite element calculations.

W. Haussmann, Approximation of subharmonic functions by harmonic ones.

D.M. Hough, Jacobi polynomial approximations to the solution of first kind integral equations in conformal mapping.

N. Jacob, Gårding's inequality and generalized Dirichlet problems for non-elliptic differential operators.

D. Kershaw, 1. A problem in axisymmetric potential theory.
2. Uniform approximation by harmonic functions.

W.A. Light, Projections in tensor product spaces.

R. Löhner, Multigrid methods for high speed compressible flow problems.

J.C. Mason, Isoparametric element nodes appropriate to square-root type singularities.

G.P. McKeown, Parallelism and approximation algorithms.

K.W. Morton, Optimal and near optimal Petrov-Galerkin methods for diffusion-convection problems.

M.J.D. Powell, Least squares data fitting using the signs of first and second order divided differences.

J.D. Pryce, A relative error measure for vectors and some applications.

A. Sharma, Convergence of complete spline interpolation for holomorphic functions.

R. Stenberg, Some mixed finite element methods for the linear elasticity problem.

P.R. Turner, Overflow, underflow and loss of significance in floating-point addition and subtraction.

J.R. Whiteman, Finite element treatment of singularities in elliptic boundary value problems.

K. Wright, A multigrid-type iterative algorithm for solving collocation equations for ordinary differential equations.

CONTENTS

MULTI-GRID CONVERGENCE THEORY *

W. Hackbusch

Mathematisches Institut, Ruhr-Universität Bochum,
Postfach 102148, D-4630 Bochum 1, Germany

Contents.

* see Lecture Notes in Mathematics, vol. 960, pp. 177 - 219

GUIDE TO MULTIGRID DEVELOPMENT **

Achi Brandt*

Department of Applied Mathematics
The Weizmann Institute of Science
Rehovot, Israel 76100

CONTENTS

---

*This research is sponsered by the Air Force Wright Aeronautical Laboratories, Air Force Systems Command, United States Air Force, under Grant AFOSR 82-0063.

** see Lecture Notes in Mathematics, vol. 960, pp. 220 - 312
(latest version available from:
GMD, Postfach 1240, D-5205 St. Augustin, Federal Republic of Germany)

FIVE LECTURES ON THE ALGORITHMIC ASPECTS OF APPROXIMATION THEORY

E. W. Cheney
Mathematics Department
The University of Texas
Austin, Texas 78712

I. Generalized Rational Approximation

In these five lectures, I will discuss several topics which illustrate the research that has been carried out in the past few years on algorithms for approximation.

There has been a revival of interest in algorithms for rational approximation, stimulated by the discovery of Barrodale, Powell, and Roberts [1] that the "ODC" Algorithm (Original Differential Correction Algorithm) is quadratically convergent under suitable hypotheses. This lecture will be devoted to their result and to subsequent additions to the theory, especially by Dua and Loeb [2].

The ODC Algorithm was proposed in [3]. Later versions [4],[7] have a simpler and more general theory, but lack the quadratic convergence. Thus the ODC Algorithm is certainly superior in most concrete computing situations. The Remez algorithm is in turn faster than ODC when the conditions are favorable, but the Remez algorithm is less robust; i.e., it is prone to failure in difficult cases. Numerical experiments have been reported in [1,5].

The setting for the ODC Algorithm is in a space $C(S)$ of all continuous real functions on a compact Hausdorff space, S. In particular, S can be a subset of $\mathbb{R}^1, \mathbb{R}^2, \ldots$; thus the theory encompasses approximation by rational functions of several variables. (This too sets ODC apart from the Remez algorithm.) In $C(S)$ the usual norm is employed, viz., $\|f\| = \sup\{|f(s)|: s \in S\}$.

We assume that two nonzero closed subspaces G and H have been prescribed. These need not be finite-dimensional, although they usually are in applications, and some of the theorems proved later will require finite-dimensionality. About H we assume that the set

$$H^+ = \{h \in H: h > 0, \|h\| = 1\}$$

is nonempty. We write $h > 0$ to mean $h(s) > 0$ for all $s \in S$. Since S is compact, this implies $\inf_s h(s) > 0$. A set of "generalized rational functions" is now defined by the equation

$$R = \{g/h: g \in G, h \in H^+\} .$$

For each f, the distance from f to R is given by

$$\operatorname{dist}(f,R) = \inf\{\|f-r\|: r \in R\} .$$

Fixing $f \in C(S)$, we pose the problem of determining a "best approximation to f in R". If it exists, it is an element $r \in R$ such that $\|f-r\| = \operatorname{dist}(f,R)$. A more modest goal is to determine a "minimizing sequence" in R. That means a sequence $\{r_k\}$ in R such that $\lim\|f-r_k\| = \operatorname{dist}(f,R)$.

<u>The ODC Algorithm</u>. Let $f \in C(S)$ and $r_0 \in R$. At step k in the algorithm an element $r_k \in R$ is available from the preceding step. Let $\Delta_k = \|f-r_k\|$ and $r_k = g_k/h_k$, where $g_k \in G$ and $h_k \in H^+$. Define a function $\delta_k\colon G \times H \to \mathbb{R}$ by the equation

$$\delta_k(g,h) = \sup_{s\in S}[\,|f(s)h(s)-g(s)| - \Delta_k h(s)]/h_k(s) .$$

Select g_{k+1} and h_{k+1} to minimize δ_k under the constraint $\|h_{k+1}\| \leq 1$. If $\delta_k(g_{k+1}, h_{k+1}) \geq 0$, stop. Otherwise continue. Put $r_{k+1} = g_{k+1}/h_{k+1}$. Define $\Delta^* = \operatorname{dist}(f,R)$.

<u>LEMMA 1</u>. <u>If</u> $h_k > 0$ <u>and</u> $\Delta_k > \Delta^*$, <u>then</u> $h_{k+1} > 0$ <u>and</u> $\Delta_{k+1} < \Delta_k$.

<u>PROOF</u>. Assume the hypotheses. Then there exists an $r \in R$ such that $\|f-r\| < \Delta_k$. Put $r = g/h$ with $g \in G$ and $h \in H^+$. Let $\alpha = \inf_s h(s)$. By the definition of (g_{k+1}, h_{k+1}), we have

$$\delta_k(g_{k+1}, h_{k+1}) \leq \delta_k(g,h) .$$

Using the definition of δ_k, we deduce that

$$\begin{aligned}\sup_s\{|fh_{k+1}-g_{k+1}| - \Delta_k h_{k+1}\}/h_k &\leq \sup_s\{|fh-g| - \Delta_k h\}/h_k \\ &= \sup_s\{|f-r| - \Delta_k\}h/h_k \leq (\|f-r\|-\Delta_k)\alpha < 0 . \qquad (*)\end{aligned}$$

(Note here that $h/h_k \geq \alpha/1$, while $\|f-r\| - \Delta_k$ is negative.) The following pointwise inequality therefore holds:

$$|fh_{k+1}-g_{k+1}| - \Delta_k h_{k+1} \leq (\|f-r\| - \Delta_k)\alpha h_k < 0 .$$

This establishes that $h_{k+1} > 0$. Dividing by h_{k+1} produces

$$|f-r_{k+1}| - \Delta_k \le (\|f-r\| - \Delta_k)\alpha h_k/h_{k+1} .$$

Taking a supremum on s yields

$$\Delta_{k+1} - \Delta_k \le -\alpha(\Delta_k - \|f-r\|)\inf h_k(s) . \quad ■$$

<u>LEMMA 2</u>. <u>If</u> $h_k > 0$ <u>and</u> $\delta_k(g_{k+1}, h_{k+1}) \ge 0$, <u>then</u> r_k <u>is a solution</u>; <u>i.e.</u> $\|f-r_k\| = \Delta^*$.

PROOF. From the inequality (*) in the preceding proof we see that if $\Delta_k > \Delta^*$ and $h_k > 0$ then

$$\delta_k(g_{k+1}, h_{k+1}) \le (\|f-r\| - \Delta_k)\alpha < 0 . \quad ■$$

<u>LEMMA 3</u>. <u>If</u> $h_k > 0$ <u>and</u> $\Delta_k = \mathrm{dist}(f,R)$, <u>then</u> $\delta_k(g_{k+1}, h_{k+1}) = 0$.

<u>PROOF</u>. From the definition of δ_k, we have $\delta_k(g_k, h_k) = 0$. Therefore $\delta_k(g_{k+1}, h_{k+1}) \le 0$. Suppose that $\delta_k(g_{k+1}, h_{k+1}) < 0$. Then we have the pointwise inequality

$$\{|fh_{k+1} - g_{k+1}| - \Delta_k h_{k+1}\}/h_k < 0 .$$

From this it follows that $|fh_{k+1} - g_{k+1}| < \Delta_k h_{k+1}$. This shows that $h_{k+1} > 0$ and that $|f-r| < \Delta_k$, which is a contradiction. ■

<u>THEOREM 1</u>. <u>If the denominators</u> h_k <u>satisfy</u> $h_k \ge \beta > 0$ <u>then</u> $\Delta_k \downarrow \Delta^*$.

<u>PROOF</u>. Let $a > \Delta^*$. We prove the lemma by showing that eventually $\Delta_k < a$. Select $r \in R$ so that $\|f-r\| < a$. Put $\lambda = \|f-r\|$. Let $r = g/h$ with $g \in G$, $h \in H^+$. Put $\alpha = \inf_s h(s)$. By the last inequality in the proof of Lemma 1,

$$\Delta_{k+1} \le \Delta_k - \alpha\beta(\Delta_k - \lambda) = (1-\alpha\beta)\Delta_k + \alpha\beta\lambda .$$

Also by Lemma 1, $h_{k+1} \in H^+$, and so $r_{k+1} \in R$. If $\Delta_{k+1} \le \lambda$, we are finished since $a > \lambda = \Delta_k \ge \Delta_{k+1} \ge \dots$. If $\Delta_{k+1} > \lambda$, then the above argument can be repeated to yield

$$\begin{aligned}\Delta_{k+2} &\le (1-\alpha\beta)\Delta_{k+1} + \alpha\beta\lambda \\ &\le (1-\alpha\beta)^2\Delta_k + [1 + (1-\alpha\beta)]\alpha\beta\lambda .\end{aligned}$$

In general, as long as $\Delta_{k+i} > \lambda$, this argument can be continued. Thus: either $\Delta_{k+i} \leq \lambda < a$ for some i, or for all i,

$$\Delta_{k+i} \leq (1-\alpha\beta)^i \Delta_k + \alpha\beta\lambda \sum_{j=0}^{i-1} (1-\alpha\beta)^j .$$

Letting $i \to \infty$, we obtain as limiting value on the right the quantity $\alpha\beta\lambda[1-(1-\alpha\beta)]^{-1} = \lambda$. Thus for all sufficiently large i we have $\Delta_{k+i} < a$. ■

THEOREM 2. Under the hypothesis $h_k \geq \beta > 0$, f has a best approximation in R, and moreover, it has one, g^*/h^*, such that $\beta \leq h^* \leq 1$.

PROOF. Theorem 1 asserts that $\Delta_k \downarrow \Delta^*$. Also $\beta \leq h_k \leq 1$. Since $\|f-g_k/h_k\| = \Delta_k \leq \Delta_0$ the sequence $\{g_k\}$ is bounded in G. The sequence $\{(g_k, h_k)\}$ in $G \times H$ therefore has a cluster point (g^*, h^*), and of course $g^* \in G$, $h^* \in H$, $\beta \leq h^* \leq 1$. Since $\|f-g_k/h_k\| = \Delta_k$ we can let $k \to \infty$ through a suitable sequence of integers and draw the conclusion that $\|f-g^*/h^*\| = \Delta^*$. ■

Because of limitations of space, the remaining results will be quoted without proof.

During the conference, Professor M.J.D. Powell kindly showed me how to improve some of my preliminary results. The following three theorems are due to Powell.

THEOREM 3. If $h_k \geq \beta > 0$ for all k then Δ_k converges at least linearly to Δ^*.

THEOREM 4. If the sequence $\{h_k\}$ converges to an element $h^* > 0$, then Δ_k converges Q-superlinearly to Δ^*.

THEOREM 5. If S is a finite set, then $\Delta_k \downarrow \Delta^*$.

Turning now to the "classical rational functions", we let $S = [a,b]$, $G = \Pi_n$, and $H = \Pi_m$ (spaces of polynomials). Then R^n_m is the set of quotients g/h with $g \in \Pi_n$, $h \in \Pi_m$, and $h > 0$ on $[a,b]$. A function f is said to be "normal with respect to R^n_m" if $\text{dist}(f,R^n_m) < \text{dist}(f,R^{n-1}_{m-1})$.

THEOREM 6. (Dua and Loeb) If f is normal with respect to R^n_m and if the ODC algorithm is started with an r_0 such that $\|f-r_0\| \leq \text{dist}(f,R^{n-1}_{m-1})$ then the denominators h_k satisfy $\inf_s \inf_k h_k(s) > 0$, and Δ_k converges quadratically to Δ^*.

LEMMA 4. (Barrodale, Powell, Roberts) If $\bar{r} \in R^n_m \setminus R^{n-1}_{m-1}$ and $\bar{r} = \bar{g}/\bar{h}$ with $\bar{g} \in \Pi_n$, $\bar{h} \in \Pi_m$, $\bar{h} > 0$, and $\|\bar{h}\| = 1$, then there is a θ such that the conditions $g \in \Pi_n$, $h \in \Pi_m$, $h > 0$, $\|h\| = 1$ imply that $\|h-\bar{h}\| \leq \theta\|g/h-\bar{r}\|$.

THEOREM 7. (Barrodale, Powell, Roberts) *If* S *is a finite subset of* $\mathbb{R}$ *containing at least* $n+m+1$ *points, if* f *is normal with respect to* R^n_m, *and if* f *has a best approximation in* R^n_m *then* Δ_k *converges quadratically to* Δ^*.

References

1. I. Barrodale, M.J.D. Powell, and F.D.K. Roberts, "The differential correction algorithm for rational approximation", SIAM J. Numer. Analysis 9(1972), 493-504.

2. S.N. Dua and H.L. Loeb, "Further remarks on the differential correction algorithm", SIAM J. Numer. Analysis 10(1973), 123-126.

3. E.W. Cheney and H.L. Loeb, "Two new algorithms for rational approximation", Numer. Math. 3(1961), 72-75.

4. E.W. Cheney and H.L. Loeb, "On rational Chebyshev approximation", Numer. Math. 4(1962), 124-127.

5. C.M. Lee and F.D.K. Roberts, "A comparision of algorithms for rational ℓ_∞ approximation", Math. Comp. 27(1973), 111-121.

6. E.H. Kaufman, S.F. McCormick, and G.D. Taylor, "An adaptive differential correction algorithm", J. Approx. Theory 37(1983), 197-211.

7. E.W. Cheney, "Introduction to Approximation Theory", Chelsea Publishing Co., New York, 1981.

II. The Alternating Algorithm in Uniformly Convex Spaces

In 1933, von Neumann defined an algorithm (the "alternating" algorithm) for constructing the orthogonal projection from a Hilbert space onto the closure of the vector sum of two subspaces. This procedure solves, for Hilbert space, the following problem. Given two subspaces, U and V, in a Banach space X, and given two "proximity maps" A and B of X onto U and V respectively, it is required to construct a proximity map onto $\overline{U+V}$.

We recall the definition of a proximity map $A: X \to U$. It is a (generally nonlinear) operator such that for each $x \in X$,

$$\|x-Ax\| = \text{dist}(x,U) = \inf\{\|x-u\|: u \in U\} .$$

In attempting to construct a best approximation of x_0, say, in $\overline{U+V}$ it is natural to try the following iterative procedure: we compute $x_1 = x_0 - Bx_0$, $x_2 = x_1 - Ax_1$, $x_3 = x_2 - Bx_2$, and so on. It is clear that $\|x_0\| \geq \|x_1\| \geq \|x_2\| \geq \cdots$ and that $x_0 - x_n \in U+V$. Von Neumann showed that in Hilbert space, $\lim_{n\to\infty}(x_0-x_n)$ is the best approximation of x_0 in $\overline{U+V}$.

Observe that $x_2 = x_1 - Ax_1 = (I-A)x_1 = (I-A)(I-B)x_0$. Thus if we put $E = (I-A)(I-B)$, then $x_{2n} = Ex_{2n-2}$, and $x_{2n} = E^n x_0$. The algorithm is the familiar process of "functional iteration" using the operator E.

LEMMA 1. If the space X is uniformly convex then in the alternating algorithm $\lim_n (x_n - x_{n+1}) = 0$.

PROOF. Suppose n is even; the other case is similar. We have

$$\|x_{n+1}\| = \|x_n - Bx_n\| = \text{dist}(x_n, V) \le \|x_n - \tfrac{1}{2} Bx_n\|$$
$$= \tfrac{1}{2}\|x_n + (x_n - Bx_n)\| = \tfrac{1}{2}\|x_n + x_{n+1}\| \le \tfrac{1}{2}(\|x_n\| + \|x_{n+1}\|) \le \|x_n\| .$$

Now if $\|x_n\| \downarrow 0$, our conclusion is obviously true. If $\|x_n\| \downarrow r > 0$, then $\frac{1}{2}\|x_n + x_{n+1}\| \to r$ also. By uniform convexity, $\|x_n - x_{n+1}\| \to 0$.

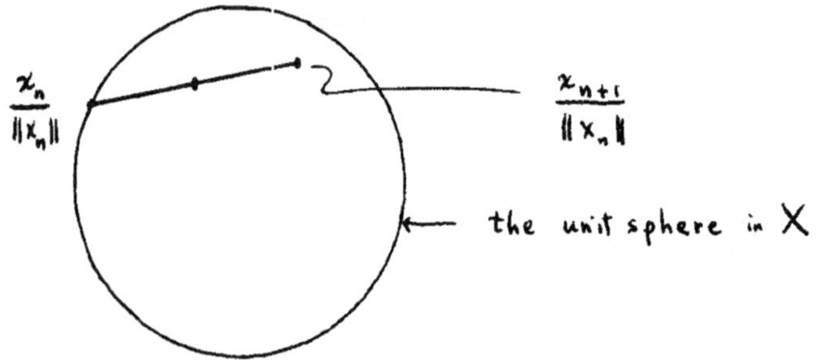

THEOREM. (Baillon, Bruck, Reich) Let K be a closed convex symmetric set in a uniformly convex Banach space. Let T be a map of K into K which is odd and nonexpansive. If $\lim_n (T^n\xi - T^{n+1}\xi) = 0$ then $\lim T^n\xi$ exists and is a fixed point of T.

LEMMA 2. If x is a fixed point of E and a smooth point of X then $\|x\| = \text{dist}(x, U+V)$.

PROOF. If $x = Ex$ then $x = x - Bx - A(x-Bx)$. Hence $Bx = -A(x-Bx)$. This shows that $Bx \in U \cap V$. Therefore

$$\|x\| = \|(x-Bx) - A(x-Bx)\| = \text{dist}(x-Bx, U) = \text{dist}(x, U) .$$

We also have

$$\|x\| = \text{dist}(x-Bx, U) \le \|x-Bx\| = \text{dist}(x, V) \le \|x\| .$$

By the Hahn-Banach Theorem, there exist two functionals $\varphi \in U^\perp$ and $\psi \in V^\perp$ such that $\|\varphi\| = \|\psi\| = 1$ and $\varphi(x) = \psi(x) = \|x\|$. Since x is a smooth point, there cannot exist two such functionals which are different from each other. Hence $\varphi = \psi$, and $\varphi \in U^{\perp} \cap V^{\perp} = (U+V)^{\perp}$. Then we see that $\|x\| = \mathrm{dist}(x, U+V)$ because if $w \in U+V$ then

$$\|x\| = \varphi(x) = \varphi(x-w) \leq \|x-w\| . \quad ■$$

THEOREM. (Light and Franchetti) Let X be a smooth and uniformly convex Banach space. Let $A: X \twoheadrightarrow U$ and $B: X \twoheadrightarrow V$ be proximity maps such that $I-A$ and $I-B$ are nonexpansive. Then $\lim(I-E^n)$ is the proximity map for $\overline{U+V}$.

PROOF. Because of the uniform convexity, each closed subspace in X has a uniquely defined proximity map. The maps A and B are odd; so are $I-A$, $I-B$, and $E \equiv (I-A)(I-B)$. By Lemma 1, $\|x_n - x_{n+1}\| \to 0$. Hence $\|x_{2n} - x_{2n+2}\| \to 0$, or $\|E^n x_0 - E^{n+1} x_0\| \to 0$. The hypotheses in the theorem of Bruck, Baillon, and Reich are therefore fulfilled, and we conclude that $E^n x_0$ converges to a fixed point of E, say x^*. Each iterate x_n in the algorithm is of the form $x_n = x_0 - w_n$ for some w_n in $U+V$. Thus $x_0 - x_n \in U+V$ and $x_0 - x^* \in \overline{U+V}$. By Lemma 2, $x_0 - x^*$ is the best approximation of x_0 in $\overline{U+V}$ since

$$\|x_0 - (x_0 - x^*)\| = \|x^*\| = \mathrm{dist}(x^*, U+V) = \mathrm{dist}(x_0, U+V) . \quad ■$$

COROLLARY. (von Neumann) In Hilbert space the alternating algorithm is effective for any pair of closed subspaces.

References

1. von Neumann, "Functional Operators", vol. II. Princeton 1950.

2. C. Franchetti and W.A. Light, "On the von Neumann algorithm in Hilbert space", Texas A&M University Report 32 (1982).

3. F. Deutsch, "The alternating method of von Neumann", in "Multivariate Approximation Theory", ed. by W. Schempp and K. Zeller, Birkhauser 1979.

4. J.B. Baillon, R.E. Bruck, and S. Reich, "On the asymptotic behavior of nonexpansive mappings and semigroups in Banach spaces", Houston J. Math. 4(1978), 1-9, MR57 #13590.

III. The Alternating Algorithm in Non-Uniformly Convex Spaces

In 1950 the Rand Corporation of Santa Monica offered a prize for the solution of a certain approximation-theoretic problem. Somewhat reformulated, the problem was to determine how well a continuous function on $[0,1]\times[0,1]$ could be approximated by a sum of two functions of one variable. Schematically, it is the approxi-problem

$$z(s,t) \approx x(s) + y(t) \ .$$

Ernst Straus and Steven Diliberto solved this problem, published the results in volume 1 of the Pacific Journal, and collected the prize. Straus told me (many years later) that the prize enabled him to attend the next International Congress of Mathematicians, but he was unable to enlighten me as to the source of the problem within the Rand Corporation; he presumed that it was related somehow to the business of killing people.

The algorithm which Diliberto and Straus devised is deservedly called by their name, although we realize that it is the alternating algorithm in a setting quite different from Hilbert space. Indeed, very little of the theory of this algorithm in uniformly convex spaces can be exploited in the setting of the space $C(S\times T)$. Furthermore, the algorithm in $C(S\times T)$ does not work on pairs of subspaces U,V in general; rather the situation is that the algorithm is effective in several isolated special cases which happen to be of some practical importance.

The problem we pose is this: given two compact Hausdorff spaces, S and T, and an element $z \in C(S\times T)$, we seek a best approximation to z by $x+y$, where $x \in C(S)$ and $y \in C(T)$.

A practical problem in numerical analysis which leads exactly to this question in approximation theory is that of optimally scaling a matrix. Let (a_{ij}) be an $m\times n$ matrix of positive elements. A "scaled copy" of (a_{ij}) is a matrix (b_{ij}) whose relation to (a_{ij}) is given by

$$b_{ij} = a_{ij}/r_i c_j$$

where r_i and c_j are positive "scale factors". Clearly we have divided row i by a row factor r_i and column j by a column factor c_j. If we are solving a system of equations with coefficient matrix (a_{ij}), each equation has been divided by a number and the variables have been replaced by multiples of them. The objective of scaling is to make the quantity $\max_{ij} b_{ij}/\min_{ij} b_{ij}$ as small as possible. If the minimum of this expression, considering all possible scalings, is λ^2, then we want to determine r_i and c_j so that $\max_{ij} b_{ij}/\min_{ij} b_{ij} = \lambda^2$. Because of homogeneity, there is nothing lost by setting $\min_{ij} b_{ij} = 1/\lambda$. Then we require

$$1/\lambda \le b_{ij} \le \lambda, \quad \text{or} \quad 1/\lambda \le a_{ij}/r_i c_j \le \lambda .$$

Taking logarithms, and putting $\varepsilon = \log \lambda$, we arrive at

$$-\varepsilon \le A_{ij} - R_i - C_j \le \varepsilon$$

where $A_{ij} = \log a_{ij}$ and so forth. Thus our problem is to approximate the bivariate function $A(i,j)$ by a sum $R(i)+C(j)$ in the supremum norm. The Diliberto-Straus algorithm is very effective on this problem.

Eight years after the appearance of the paper by Diliberto and Straus, Michael Golomb re-examined the algorithm and extracted from the original analysis the hypotheses in Banach space terms which made the algorithm work. We present Golomb's theory first.

DEFINITION. Let $A: X \twoheadrightarrow U$ be a proximity map from a Banach space X onto a subspace U. If

$$\|x-Ax+u\| = \|x-Ax-u\|$$

for all $x \in X$ and $u \in U$, then A is called a central proximity map.

EXAMPLE. Orthogonal projections in Hilbert space are central proximity maps. Indeed $x - Ax \perp U$, so $\|x-Ax \pm u\|^2 = \|x-Ax\|^2 + \|u\|^2$ by Pythagoras' Theorem.

THEOREM. (Golomb) Let U and V be subspaces of a Banach space having central proximity maps. If $U+V$ is closed, then the sequence $\{x_n\}$ generated by the alnating algorithm has the property $\|x_n\| \downarrow \operatorname{dist}(x_0, U+V)$.

This theorem is rather deep. I recommend the exposition of it in Chapter IV of the forthcoming volume [4]. An easy corollary, which includes the theorem of von Neumann, is:

THEOREM. If X is uniformly convex and if A and B are central proximity maps, then the alternating algorithm produces a sequence $\{x_n\}$ such that $\lim(x_0-x_n)$ is the best approximation of x_0 in $\overline{U+V}$.

PROOF. We have $x_0 - x_n \in U+V$ as usual, and $\|x_n\| \downarrow \operatorname{dist}(x_0, U+V)$ by Golomb's Theorem. Hence $x_0 - x_n$ is a minimizing sequence (for x_0) in $U+V$. By the uniform convexity, $\lim(x_0 - x_n)$ exists and is the best approximation of x_0 in $\overline{U+V}$. ∎

We turn our attention now to the problem originally addressed by Diliberto and Straus. Here $X = C(S \times T)$, $U = C(S)$, and $V = C(T)$. There are many proximity maps, but the natural ones are

$$(Az)(s,t) = \frac{1}{2}\max_t z(s,t) + \frac{1}{2}\min_t z(s,t)$$

$$(Bz)(s,t) = \frac{1}{2}\max_s z(s,t) + \frac{1}{2}\min_s z(s,t) .$$

Note that although we write $(Az)(s,t)$, in fact there is no dependence on t, and so $Az \in C(S)$. Likewise $Bz \in C(T)$. The univariate operator

$$\mathcal{M}f = \frac{1}{2}\max_s f(s) + \frac{1}{2}\min_s f(s) \qquad\qquad f \in C(S)$$

produces constants of best approximation to every f. See sketch.

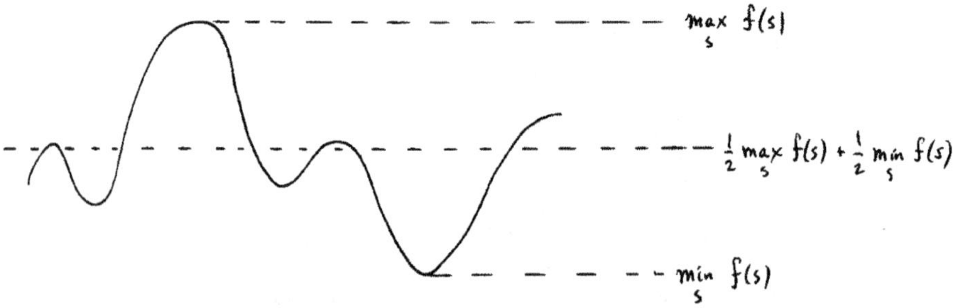

If we define the sections of a bivariate function z by

$$z_s(t) = z(s,t) = z^t(s)$$

then we see that $(Az)_s = \mathcal{M}z_s$. Thus for each s, $(Az)_s$ is the constant of best approximation to the section z_s. Here are the crucial results:

LEMMA. The proximity maps A and B defined above are nonexpansive central proximity maps.

LEMMA. (Aumann) Let $\{z_n\}$ be the sequence produced by applying the Diliberto-Straus algorithm to $z_0 \in C(S \times T)$. Then $z_{n+1} - z_n \to 0$.

THEOREM. $\{z_0 - z_n\}$ converges uniformly to a best approximation of z_0 in $C(S) + C(T)$.

The analogous problem in $L_2(S \times T)$ can be solved by the same method (the alternating algorithm), but only two steps are required to arrive at the solution. The reason for this is that the two proximity maps in this case will be the orthogonal projections $A: L_2(S \times T) \twoheadrightarrow L_2(S)$ and $B: L_2(S \times T) \twoheadrightarrow L_2(T)$. These are, of course, linear, and satisfy $\|I-A\| = \|I-B\| = 1$. Moreover, they commute: $AB = BA$. Then the theory of Boolean sums tells us that $A \oplus B$ (i.e. $A + B - AB$) is a projection of $L_2(S \times T)$ onto $L_2(S) + L_2(T)$. One verifies that $A \oplus B$ is the orthogonal projection. Two steps in the alternating algorithm produce $z_2 = z_0 - Az_0 - B(z_0 - Az_0)$ whence $z_0 - z_2 = (A \oplus B)z_0$.

THEOREM. If A and B are linear proximity maps defined on a Banach space X, and if $ABA = BA$, then $A \oplus B$ is a linear proximity map of X onto the vector sum of the ranges of A and B [5].

References

1. S. Diliberto and E. Straus, "On the approximation of a function of several variables by the sum of functions of fewer variables", Pacific J. Math. 1(1951), 195-210. MR13-334.

2. M. Golomb, "Approximation by functions of fewer variables" in "On Numerical Approximation", R.E. Langer, editor. University of Wisconsin Press 1959, pp. 275-327. MR21#962.

3. G. Aumann, "Uber approximative Nomographie. I". Bayer. Akad. Wiss. Math.-Nat. Kl. S.-B. 1958, 137-155. MR22#1101. Part II, ibid, 1959, 103-109. MR22#6968. Part III, ibid, 1960, 27-34. MR24#B1289.

4. W. A. Light and E. W. Cheney, "Approximation Theory in Tensor Product Spaces", Lecture Notes in Mathematics, Springer-Verlag, New York. To appear.

5. J. R. Respess and E. W. Cheney, "Best approximation problems in tensor product spaces", Pacific J. Math. 102(1982), 437-446.

IV. Minimal Projections

A projection of a Banach space X onto a subspace Y is a bounded linear operator $P: X \twoheadrightarrow Y$ such that $Py = y$ for all $y \in Y$. Because of the elementary inequalities

$$\|x-Px\| \leq \|I-P\| \cdot \mathrm{dist}(x,Y) \leq (1 + \|P\|)\mathrm{dist}(x,Y)$$

projections of small norm produce good approximations. They are often used as a substitute for the (generally nonlinear) proximity maps which have the property

$$\|x-Ax\| = \mathrm{dist}(x,Y) .$$

A notable example is the projection $P: C[-1,1] \twoheadrightarrow \Pi_n$ which gives a Lagrange interpolant at nodes which are the zeros of the Chebyshev polynomial T_{n+1}.

We pose the problem of finding minimal projections.

Minimal Projection Problem. Given a Banach space X and a subspace Y in X, find the projections of least norm from X onto Y.

In this context, it is helpful to give a name to the minimal norm which we wish to achieve. It is called the relative projection constant of Y in X and is defined by

$$\lambda(Y,X) = \inf\{\|P\|: P \text{ is a projection of } X \text{ onto } Y\} .$$

If Y is finite dimensional, the important inequality of Kadec and Snobar applies. According to it,

$$\lambda(Y,X) = \sqrt{\dim(Y)} .$$

On the other hand, for some pairs (Y,X), we have

$$\lambda(Y,X) = +\infty .$$

Thus there exists in such a case no (bounded linear) projection of X onto Y.

If Y is an n-dimensional subspace in X, then let us select a basis $\{y_1, \ldots, y_n\}$ for Y. Let $P: X \twoheadrightarrow Y$ be a projection. For each x, Px must be a linear combination of $y_1, \ldots, y_n$, for example, $Px = \sum_{i=1}^{n} \lambda_i y_i$. Since the coefficients λ_i depend on x we write

$$Px = \sum_{i=1}^{n} \lambda_i(x)\cdot y_i .$$

Elementary arguments show that the λ_i are continuous and linear; i.e., $\lambda_i \in X^*$. The equation $Py_j = y_j$ leads at once to the requirement $\lambda_i(y_j) = \delta_{ij}$. Except for details, we have proved:

THEOREM. Let Y be an n-dimensional subspace with basis $\{y_1, \ldots, y_n\}$ in a Banach space X. The projections of X onto Y are in 1-1 affine correspondence with the n-tuples $(\lambda_1, \ldots, \lambda_n)$ where $\lambda_i \in X^*$ and $\lambda_i(y_j) = \delta_{ij}$.

This simple result shows that the set of all projections from X onto Y is an immense linear manifold in the space $\mathcal{L}(X,Y)$ of all bounded linear operators from X into Y.

About the numerical determination of minimal projections very little is known except in spaces of continuous functions. Some of the theory involved here will now be outlined. Suppose then that T is a compact separable Hausdorff space. Thus T contains a dense sequence $\{t_1, t_2, \ldots\}$. As usual $C(T)$ denotes the Banach space of all continuous functions on T normed by

$$\|x\| = \max\{|x(t)|: t \in T\} .$$

For discretization purposes it is convenient to introduce the seminorms

$$\|x\|_k = \max\{|x(t_1)|, |x(t_2)|, \ldots, |x(t_k)|\} .$$

Since the sequence $\{t_i\}$ is dense, we have $\|x\|_k \uparrow \|x\|$ for all $x \in C(T)$. If L is a linear operator on $C(T)$, its standard norm is

$$\|L\| = \sup\{\|Lx\|: x \in C(S) \,\&\, \|x\| \leq 1\} .$$

We also have some norm-like functions

$$\|L\|_k = \sup\{\|Lx\|_k: x \in C(S) \,\&\, \|x\|_k \leq 1\} .$$

The problem that <u>can</u> be solved is to determine a projection P, supported on $\{t_1, \ldots, t_m\}$, mapping $C(T)$ onto Y, and having $\|P\|_k$ a minimum. We assume that $\dim(Y) = n \leq m \leq k$. To say that P is supported on $\{t_1, \ldots, t_m\}$ means that each functional λ_i appearing in the equation $Px = \sum_{i=1}^{n} \lambda_i(x) y_i$ is a linear combination of the point functionals $\hat{t}_1, \ldots, \hat{t}_m$. Thus

$$\lambda_i = \sum_{j=1}^{m} c_{ij} \hat{t}_j \qquad 1 \leq i \leq n$$

or

$$\lambda_i(x) = \sum_{j=1}^{m} c_{ij} x(t_j) \qquad x \in C(T), \quad 1 \leq i \leq n .$$

(The term "support" is borrowed from measure theory.)

Now let $P_k^{(m)}$ be a projection of minimal norm $\|P_k^{(m)}\|_k$ supported on $\{t_1, \ldots, t_m\}$. We state an important theorem concerning the discretization process.

<u>Discretization Theorem</u>. <u>The sequence</u> $\{P_m^{(m)}\}$ <u>has cluster points in the norm topology of</u> $\mathcal{L}(X,Y)$. <u>Each of these is a minimal projection of</u> $C(T)$ <u>onto</u> Y. <u>Furthermore</u> $\lim\|P_m^{(m)}\|$ <u>exists and is the projection constant of</u> Y <u>in</u> $C(T)$.

For the proof of this and other matters discussed here see [1] and [3].

<u>Computation of</u> $P_k^{(m)}$. Since $P_k^{(m)}$ is supported on $\{t_1,\ldots,t_m\}$ it has a representation in the form

$$P_k^{(m)}x = \sum_{i=1}^{m} x(t_i)w_i \qquad\qquad x \in C(S)$$

where w_i are appropriate elements of Y, not independent in general. The k-norm of $P_k^{(m)}$ is

$$\begin{aligned}\|P_k^{(m)}\|_k &= \sup_{\|x\|_k\le 1} \|P_k^{(m)}x\|_k \\ &= \sup_{\|x\|_k\le 1}\ \max_{1\le j\le k} |(P_k^{(m)}x)(t_j)| \\ &= \sup_{\|x\|_k\le 1}\ \max_{1\le j\le k} \left|\sum_{i=1}^{m} x(t_i)w_i(t_j)\right| \\ &= \max_{1\le j\le k} \sum_{i=1}^{m} |w_i(t_j)| \ .\end{aligned}$$

The preceding equation indicates what is to be minimized, viz., $\max_{1\le j\le k} \sum_{i=1}^{m} |w_i(t_j)|$. This is known as the <u>Lebesgue function</u>. The w_i range over Y, but must be chosen so that $P_k^{(m)}$ is a projection. This condition is that

$$P_k^{(m)}y_i = y_i \qquad\qquad (1 \le i \le n)$$

or

$$y_i = \sum_{j=1}^{m} y_i(t_j)w_j \qquad\qquad (1 \le i \le n) \ .$$

One can express each w_i as a linear combination of the basis elements $y_1,\ldots,y_n$

$$w_i = \sum_{j=1}^{n} \alpha_{ij}y_j \qquad\qquad (1 \le i \le m) \ .$$

This brings us to the first formulation of our problem.

<u>First Form of the Extremum Problem:</u> Determine an $m\times n$ matrix (α_{ij}) which minimizes the expression

$$\max_{1\le j\le k} \sum_{i=1}^{m} \left|\sum_{\nu=1}^{n} \alpha_{i\nu}y_\nu(t_j)\right|$$

subject to the constraints

$$\sum_{\nu=1}^{m} y_i(t_\nu)\alpha_{\nu j} = \delta_{ij} \qquad\qquad (1 \le i,j \le n) \ .$$

This extremal problem can be put into various other forms which are more convenient for computation. For example, we could employ a particular basis $\{y_1, \dots, y_n\}$ for which

$$y_i(t_j) = \delta_{ij} \qquad (1 \le i \le n,\ 1 \le j \le n) .$$

Then for convenience, put $y_{n+1} = y_{n+2} = \dots = 0$. Our problem can be expressed now as an unconstrained minimum problem.

<u>2nd Form of the Extremum Problem</u>. Here we let $k = \infty$ so that $\|f\|_k = \|f\|$. Fixing m, let S denote the subset of $\mathbb{R}^m$ consisting of points $s = (s_1, s_2, \dots, s_m)$ such that $|s_i| = 1$ for all i. On $S \times T$ we define

$$h(s,t) = \sum_{j=1}^{m} s_j y_j(t)$$

and

$$g_{\nu\mu}(s,t) = y_\nu(t) \sum_{j=1}^{m} s_j D_{\mu j} \qquad (1 \le \nu \le n,\ 1 \le \mu \le m-n)$$

where

$$D_{ij} = \begin{cases} -y_j(t_{n+i}) & \text{if } 1 \le j \le n \text{ and } 1 \le i \le m-n \\ \delta_{j,n+i} & \text{if } n+1 \le j \le m \text{ and } 1 \le i \le m-n . \end{cases}$$

Find the best approximation (in the supremum norm) of h by a linear combination of the functions $g_{\nu\mu}$. That is, find coefficients $c_{\nu\mu}$ to minimize

$$\max_{s\in S} \max_{t\in T} \left| h(s,t) - \sum_{\nu=1}^{n} \sum_{\mu=1}^{m-n} c_{\nu\mu} g_{\nu\mu}(s,t) \right| .$$

This is a standard Chebyshev approximation problem, to which the Remez algorithm can be applied. Once the coefficients $c_{\nu\mu}$ have been determined, the minimal projection supported on $\{t_1, \dots, t_m\}$ is given by

$$B = CD$$

$$Q = \sum_{i=1}^{n} \hat{t}_i \otimes y_i \qquad \text{i.e.} \qquad Qx = \sum_{i=1}^{n} x(t_i) y_i$$

$$P = Q - L$$

$$L = \sum_{i=1}^{m} \hat{t}_i \otimes u_i \qquad \text{i.e.} \qquad Lx = \sum_{i=1}^{m} x(t_i) u_i$$

$$u_i = \sum_{j=1}^{n} B_{ji} y_j .$$

Here B is an $n\times m$ matrix, C is $n\times(m-n)$, and D is $(m-n)\times m$. The details of this reduction to "second extremal form" are not given here but are in [3]. A series of numerical experiments using the Remez algorithm was carried out. A typical result yielded the projection of least norm among all projections supported on $\{-1,-4/5,-3/5,0,1/2,7/10,1\}$ from $C[-1,1]$ onto Π_2. Its norm was 1.22591. (The projection of least norm in the family of all projections in this case is not known.) Observe that in determining a minimal projection supported on m points from $C(T)$ onto an n-dimensional subspace, we must solve a Chebyshev approximation problem involving a subspace of dimension $n(m-n)$.

References

1. P.D. Morris and E.W. Cheney, "On the existence and characterization of minimal projections", J. für die Reine und Angewandte Math. 270(1974), 61-76.

2. E.W. Cheney, "Projections with finite carrier" in "Numerische Methoden der Approximationstheorie" ISNM vol. 16, Birkhäuser-Verlag, Basel, 1972, pp. 19-32. MR53#8738.

3. E.W. Cheney and P.D. Morris, "The numerical determination of projection constants", in ISNM vol. 26, Birkhäuser-Verlag, Basel 1975, pp. 29-40.

4. M.I. Kadec and M.G. Snobar, "Certain functionals on the Minkowski compactum", Math. Notes 10(1971), 694-696. MR45#861.

5. D.L. Motte, "A constructive approach to minimal projections in Banach spaces", J. Approximation Theory, to appear. (Ph.D. Thesis, University of California at Riverside, 1982.)

V. Numerical Construction of Chebyshev Centers

In practical problems of approximation it sometimes happens that instead of approximating a single well-defined function , we wish to approximate a set of functions. This would happen, for example, if a function was not known precisely but was known to belong to a certain set. The problem comes up in a natural way also when we attempt to approximate a multivariate function by a function of fewer variables.

Let S be a bounded set in a Banach space X. Let Y be a subset of X. We wish to find a single element of Y which is a good approximation to all of the elements of S simultaneously. The expression that is to be made a minimum (by choosing $y \in Y$) is therefore

$$\sup_{x\in S}\|y-x\| .$$

The minimum value which we seek is

$$r_Y(S) = \inf_{y \in Y} \sup_{x \in S} \|y-x\| .$$

This is called the Chebyshev radius of S with respect to Y.

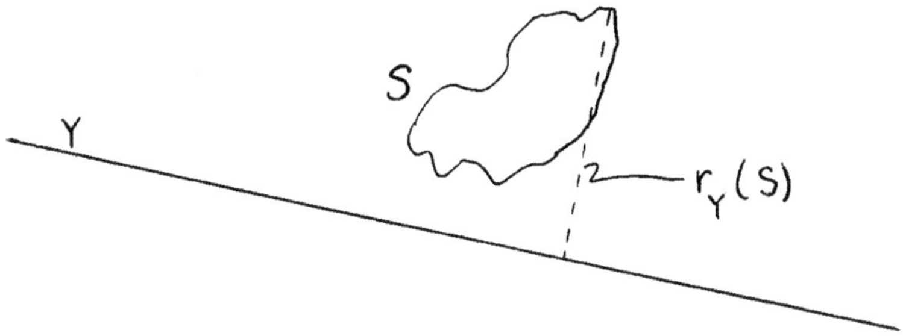

Figure 1 Chebyshev radius of S with respect to Y

The set of all solutions to our problem is called the Cheybshev center of S with respect to Y:

$$E_Y(S) = \{y \in Y: \sup_{x \in S} \|x-y\| = r_Y(S)\} .$$

The approximation problem outlined above is also called "simultaneous" approximation or "global" approximation. If Y is taken to be the entire space X, the problem is still of interest. Then the entities $r_X(S)$ and $E_X(S)$ are written simply $r(S)$ and $E(S)$. They are called just the Chebyshev radius and center of S.

The connection between Chebyshev centers and multivariate approximation is as follows. Suppose that a continuous function of two variables is to be approximated by a continuous function of one variable; that is, we have $f \in C(S \times T)$ and seek $y \in C(S)$ to minimize

$$\sup_{s \in S} \sup_{t \in T} |f(s,t)-y(s)| .$$

If we define $f^t(s) = f(s,t)$ and put

$$K = \{f^t: t \in T\}$$

then the expression to minimize is

$$\sup_{t\in T}\|f^t - y\| = \sup_{x\in K}\|x-y\|$$

and therefore y should be any element in $E(K)$.

The problem of determining an element in the Chebyshev center of S relative to Y can be recast as a problem of approximating a single entity in an appropriate space. Namely, we define the Banach space $C(S,X)$ of all bounded continuous maps from the set S into the Banach space X. Such a map receives the norm $\|f\| = \sup_{s\in S}\|f(s)\|$. Next, define, for each $y \in Y$, an element $\bar{y} \in C(S,X)$ by putting $\bar{y}(s) = y$ (a constant function). Put $\bar{Y} = \{\bar{y}: y \in Y\}$, and observe that $\bar{Y}$ is a bounded subset of $C(S,X)$. Finally, define e in $C(S,X)$ by the equation $e(s) = s$. Now if we try to approximate the single function e by an optimal element in $\bar{Y}$, we will be attempting to minimize

$$\|e-\bar{y}\| = \sup_{s\in S}\|e(s)-\bar{y}(s)\| = \sup_{s\in S}\|s-y\| .$$

Thus, simultaneous approximation of S by an element of Y is equivalent to approximating e by an element of $\bar{Y}$. Of course, the latter is in a more "exotic" space, $C(S,X)$, but these spaces form a very important category of Banach spaces, (which deserve to be more extensively studied in approximation theory).

Chebyshev centers in spaces $C(T)$ are rather well understood by now. For a general topological space T, $C(T)$ denotes the space of bounded continuous real functions on T. The following theorem was proved originally for compact T by Zamjatin [1]. See also [2] and [3].

THEOREM. *Let* T *be any topological space and* A *any bounded subset of* $C(T)$. *Then the Chebyshev center of* A *is nonempty.*

The theory of centers in $C(T)$ is simpler if we consider T compact and A compact in $C(T)$. In order to explain this case, let

$$u(t) = \inf\{a(t): a \in A\}$$
$$v(t) = \sup\{a(t): a \in A\} .$$

If A is compact, these two functions are continuous. Now let $\rho > 0$ and $y \in Y$. The following are equivalent:

(i) $\sup_{a\in A}\|y-a\| \leq \rho$

(ii) $\|y-a\| \leq \rho$ for all $a \in A$

(iii) $-\rho \leq y - a \leq \rho$ for all $a \in A$

(iv) $-\rho + a \leq y \leq \rho + a$ for all $a \in A$

(v) $-\rho + v \leq y \leq \rho + u$

(vi) $-\rho \leq y - v \leq y - u \leq \rho$

(vii) $\|y-v\| \leq \rho \qquad \|y-u\| \leq \rho$

(viii) $\max\{\|y-v\|, \|y-u\|\} \leq \rho.$

This proves the following result.

THEOREM 1. The simultaneous approximation of the compact set A is equivalent to the simultaneous approximation of the pair $\{u,v\}$.

THEOREM 2.

(a) $r(A) = \frac{1}{2}\|v-u\|$

(b) $E(A) = \{x \in C(T): v-r(A) \leq x \leq u + r(A)\}$

(c) $\frac{1}{2}(u+v) \in E(A)$.

PROOF. If $x \in C(T)$ then for any $t \in T$,

$$\begin{aligned}\sup_{a\in A}\|x-a\| &= \max\{\|x-u\|, \|x-v\|\} \\ &\geq \max\{x(t)-u(t), v(t)-x(t)\} \\ &\geq \frac{1}{2}\{x(t)-u(t)+v(t)-x(t)\} = \frac{1}{2}(v-u)(t) .\end{aligned}$$

By taking first a supremum in t and then an infimum on x, we obtain

$$r(A) \geq \frac{1}{2}\|v-u\| .$$

It is easy to verify that the element $x_0 = \frac{1}{2}(u+v)$ satisfies $\|x_0-u\| = \|x_0 - v\| = \frac{1}{2}\|v-u\|$. Hence $r(A) = \frac{1}{2}\|v-u\|$, and assertions (a) and (c) are established. The inequalities leading to Theorem 1 establish part (b). ∎

The analogue of Theorem 2 for relative Chebyshev centers was established by Smith and Ward in [7]. See also [2]. We state it for compact T and compact A although a more general version exists in [7] and [2].

THEOREM 3.

(a) $E_Y(A) = r(A) + \text{dist}(Y, E(A))$.

(b) $E_Y(A) = \{y \in Y: v-r_Y(A) \leq y \leq u + r_Y(A)\}$

(c) In order for $E_Y(A)$ to be nonempty it is necessary and sufficient that $\inf\{\text{dist}(x,Y): x \in E(A)\}$ be attained.

Discretization. Let T be a compact metric space, and let $\{t_1, t_2, \ldots\}$ be a countable dense set in T. In $C(T)$ there is a family of semi-norms defined by

$$\|f\|_k = \max\{|f(t_i)| : 1 \leq i \leq k\} \qquad k = 1,2,\ldots .$$

THEOREM. (Amir and Ziegler) Let A be a compact subset of $C(T)$ and let Y be a finite-dimensional closed convex set. For $k = 1,2,\ldots$ let y_k be a point in Y which minimizes the expression $\sup_{a \in A} \|a-y\|_k$. Then each cluster point of the sequence $\{y_k\}$ is a point in the relative Chebyshev center of A with respect to Y.

Because of this theorem, it would be desirable to have an algorithm for Chebyshev centers in $C(T)$ when T is a finite set. Amir and Ziegler describe such an algorithm for the case when Y is a finite-dimensional Haar subspace on $[a,b]$ and T is a finite subset of $[a,b]$. (A Haar subspace of dimension n has the property that 0 is the only member having n zeros in $[a,b]$.)

References

1. V.M. Zamyatin, "Chebyshev centers in the space C(S)", First Scientific Conference of Young Scholars of the Akygei, 1971, pp. 28-35.

2. C. Franchetti and E.W. Cheney, "Simultaneous approximation and restricted Chebyshev centers in function spaces", in "Approximation Theory and Applications", ed. by Z. Ziegler. Academic Press 1981.

3. R.B. Holmes, "A Course on Optimization and Best Approximation", Lecture Notes in Mathematics, vol. 257, Springer-Verlag, 1972.

4. D. Amir and Z. Ziegler, "Construction of elements of the relative Chebyshev center", in "Approximation Theory and Applications", ed. by Z. Ziegler. Academic Press 1981.

5. D. Amir and Z. Ziegler, "Relative Chebyshev centers in normed linear spaces", parts I and II. J. Approximation Theory.

6. D. Amir, J. Mach, Saatkamp, "Existence of Chebyshev centers", Trans. American Math. Soc. 271(1982), 513-524.

7. P.W. Smith and J.D. Ward, "Restricted centers in C(Ω)", Proc. Amer. Math. Soc. 48(1975), 165-172.

Typed by Nita Goldrick.

LECTURES ON OPTIMAL RECOVERY

C. A. Micchelli and T. J. Rivlin
Mathematical Sciences Department
IBM Thomas J. Watson Research Center
Yorktown Heights, N. Y. 10598
U.S.A.

Table of Contents

1. Introduction

In 1977 in a long paper (Micchelli and Rivlin [37], henceforth referred to as M-R) the authors presented a general framework and many examples of what they called "optimal recovery". Optimal recovery means estimating some required feature of a function, known to belong to some specified class of functions, from limited information about it as effectively as possible. In these lectures we propose to recall the basic notions of our approach and report on some of the advances in the area of optimal recovery not mentioned in, or subsequent to, M-R. We begin by introducing the basic concepts of optimal recovery by means of simple examples which also serve as previews of later results.

Let B be a compact subset of euclidean m-space. $L^\infty(B)$ is the space of (essentially) bounded functions (real or complex-valued) on B. The set of *contractions* in B is $K = \{f \in L^\infty(B): |f(x) - f(y)| \leq |x - y|;\ x,y \in B\}$. Suppose that we know: (i) $f \in K$, and (ii) $f(x_1),\ldots,f(x_n)$, where $x_1,\ldots,x_n$ are given distinct points of B.

Example 1.1 Given $w \in B$ what is a *best possible* estimate, based solely on the information (i) and (ii), of $f(w)$? This is the problem of optimal interpolation.

To be more precise: suppose $f \in K$ and $If := (f(x_1),\ldots,f(x_n))$. We call any function α,

$$\alpha : IK \to \mathbb{R}\ (\mathbb{C})$$

an *algorithm*.

$$E(\alpha) = \sup_{f \in K} |f(w) - \alpha(If)|$$

is the *error* of algorithm α, and

$$E^* = \inf_{\alpha} E(\alpha)$$

is the *intrinsic error* in the problem.

If $E(\alpha^*) = E^*$, α^* is an *optimal algorithm* and effects the *optimal recovery* of $f(w)$.

A useful observation in helping to solve this problem is the following:

$$(1.1) \qquad E^* \geq \sup\{|f(w)| : f \in K,\ f(x_i) = 0,\ i = 1,\ldots,n\}.$$

Proof. $f \in K \Leftrightarrow -f \in K$ and $If = 0 \Leftrightarrow I(-f) = 0$. For any α we have $|f(w) - \alpha(0)| \leq E(\alpha)$ and $|-f(w) - \alpha(0)| = |f(w) + \alpha(0)| \leq E(\alpha)$. Hence $|f(w)| \leq E(\alpha)$ and (1.1) follows.

Put

$$B_i = \{x \in B : \min_j |x - x_j| = |x - x_i|\},\ i = 1,\ldots,n.$$

Suppose $w \in B_k$. Let $q(x) = |x - x_i|$, $x \in B_i$, $i = 1,...,n$. Then $q(x_i) = 0$, $i = 1,...,n$ and $q \in K$. Hence in view of (1.1) we have

$$E^* \geq q(w) = |w - x_k|.$$

Let us define the algorithm β by

$$\beta : (f(x_1),...,f(x_n)) \to f(x_k).$$

Then

$$E(\beta) = \sup_{f \in K} |f(w) - \beta(If)| = \sup_{f \in K} |f(w) - f(x_k)| \leq |w - x_k| = q(w).$$

Thus $E(\beta) \leq q(w) \leq E^* \leq E(\beta)$. Therefore $E(\beta) = E^* = q(w)$ and β is an optimal algorithm for interpolation. Note also that equality holds in (1.1).

Example 1.2 In the same setting and with the same information recover the function f (instead of the functional f(w)) optimally. This is the problem of optimal approximation. Note that an algorithm now is any function α whose domain is IK and whose range is in $L^{\infty}(B)$.

The analogue of (1.1)

$$E^* \geq \sup\{\|f\|_{\infty} : f \in K, f(x_i) = 0, i = 1,...,n\} \tag{1.2}$$

is easily established by imitating the proof of (1.1), and we thus obtain $E^* \geq \|q\|_{\infty}$. Let s be the step-function ($\in L^{\infty}(B)$) defined by

$$s(x) = f(x_j), \; x \in \text{int } B_j, \; j = 1,...,n.$$

Consider the algorithm $\gamma : (f(x_1),...,f(x_n)) \to s$.
If $x \in \text{int } B_j$ we have, for $j = 1,...,n$,

$$|f(x) - \gamma(If)| = |f(x) - f(x_j)| \leq |x - x_j| = q(x) \leq \|q\|_{\infty}.$$

Thus $E(\gamma) \leq \|q\|_{\infty} \leq E^* \leq E(\gamma)$, hence $E(\gamma) = E^* = \|q\|_{\infty}$. γ is an optimal algorithm for our problem of approximation.

In connection with this problem it is appropriate to raise the question of optimal information, or, more particularly, optimal location of the sampling points $x_1,...,x_n$. Namely, choose $x_1,...,x_n$ ($\in B$) so as to minimize the intrinsic error, $E^*(x_1,...,x_n)$. This leads to the extremal problem

$$\inf_{x_1,\ldots,x_n} \|q\|_\infty = \inf_{x_1,\ldots,x_n} \max_{x\in B} \min_i |x - x_i| .$$

This is a minimal covering problem (for B) which does not lend itself to simple resolution for general B. We shall discuss some special cases in a moment.

Example 1.3 Optimal integration. We now seek the optimal recovery of

$$\int_B f\, dx$$

from the same information as in the previous examples. Again we have the analogue of (1.1)

$$E^* \geq \sup \{ | \int_B f\, dx | : f \in K, f(x_i) = 0, i = 1,\ldots,n\} \tag{1.3}$$

and hence

$$E^* \geq \int_B q\, dx .$$

Let λ be the algorithm defined by

$$\lambda : (f(x_1),\ldots,f(x_n)) \to \int_B s\, dx = \sum_{j=1}^n b_j f(x_j)$$

where b_j is the volume of B_j, $j = 1,\ldots,n$. Then

$$| \int_B f\, dx - \int_B s\, dx | \leq \int_B |f - s|\, dx \leq \int_B q\, dx,$$

and we conclude, as before, that λ is optimal and

$$E^* = \int_B q\, dx .$$

Note that the location of the optimal sampling points now requires the determination of points $x_1,\ldots,x_n$ for which

$$\min_{x_1,\ldots,x_n} \int_B q\, dx = \min_{x_1,\ldots,x_n} \sum_{j=1}^n \int_{B_j} |x - x_j|\, dx$$

is attained.

We turn now to some special cases. Consider optimal approximation with $m = 1$, $B = [0,1]$, $0 \leq x_1 < \ldots < x_n \leq 1$. The B_i are now intervals, namely, $B_1 = [0,\xi_1]$, $B_i = [\xi_{i-1}, \xi_i]$, $i = 2,\ldots,n-1$, $B_n = [\xi_{n-1},1]$, where

$$\xi_i = \frac{x_i + x_{i+1}}{2}, \quad i = 1,\ldots,n-1 .$$

$q(x)$ is now $|x - x_i|$ in each interval $[\xi_{i-1}, \xi_i]$, $i = 1,\ldots,n$, ($\xi_0 = 0$, $\xi_n = 1$), and so if we put $\Delta = \max (x_{i+1}-x_i)$, $i = 1,\ldots,n-1$, we obtain

$$E^* = \|q\|_\infty = \max (x_1, \frac{\Delta}{2}, 1 - x_n) .$$

An optimal approximation of f is given by the step-function $s(x) = f(x_i)$, $\xi_{i-1} < x < \xi_i$, $i = 1,\ldots,n$.

The optimal sampling points are easy to determine in this case. They are

$$x_i = \frac{2i-1}{2n}, \quad i = 1,\ldots,n$$

and the intrinsic error for this choice of points is $E^* = 1/(2n)$. Note that these same points are optimal for the optimal integration problem on [0,1], and produce an intrinsic error $1/(4n)$.

There is another simple optimal algorithm for the approximation problem, piecewise linear interpolation. Namely, let δ map the data onto the piecewise linear interpolant. I.e.,

$$\delta : (f(x_1),\ldots,f(x_n)) \to p(x) = \begin{cases} f(x_1), & 0 \leq x \leq x_1 \\ f(x_i)\dfrac{x_{i+1}-x}{x_{i+1}-x_i} + f(x_{i+1})\dfrac{x-x_i}{x_{i+1}-x_i}, & i = 1,\ldots,n-1 \\ f(x_n), & x_n \leq x \leq 1 . \end{cases}$$

(Note that $p \in K$.) Suppose that $x_i \leq x \leq x_{i+1}$. Then $x = \lambda_1(x)\, x_i + \lambda_2(x)\, x_{i+1}$, $\lambda_1, \lambda_2 \geq 0$, $\lambda_1 + \lambda_2 = 1$. $p(x) = \lambda_1\, p(x_i) + \lambda_2 p(x_{i+1})$ and

$$\begin{aligned} f(x)-p(x) &= \lambda_1\, [f(x) - p(x_i)] + \lambda_2[f(x)-p(x_{i+1})] \\ &= \lambda_1\, [f(x)-f(x_i)] + \lambda_2[f(x) - f(x_{i+1})] , \end{aligned}$$

which yields

$$|f(x)-p(x)| \le \lambda_1(x-x_i)+\lambda_2(x_{i+1}-x)=2\lambda_1\lambda_2(x_{i+1}-x_i)$$

$$\le \frac{x_{i+1}-x_i}{2} .$$

Thus for $0 \le x \le 1$ we have

$$|f(x)-p(x)| \le \max\left(x_1, \frac{\Delta}{2}, 1-x_n\right) = E^* .$$

Piecewise linear interpolation is an optimal algorithm.

We consider next optimal approximation in the case $m = 2$, $n = 3$. B is a triangle (having no obtuse angle) with vertices x_1, x_2, x_3. Let C be the circumcenter of B. $(C \in B)$. B_1, B_2, B_3 are seen to be convex quadrilaterals meeting at C. Without loss of generality we may choose coordinates so that $C = 0$ and the radius of the circumscribing circle of B is 1. Then clearly $\|q\|_\infty = 1$. Let ζ map $(f(x_1), f(x_2), f(x_3))$ onto the linear interpolant ℓ. That is $\ell(u,v)$ is linear and $\ell(u_i, v_i) = f(u_i, v_i)$, $i = 1,2,3$ where $x_i = (u_i, v_i)$, $i = 1,2,3$, and $x = (u,v)$.

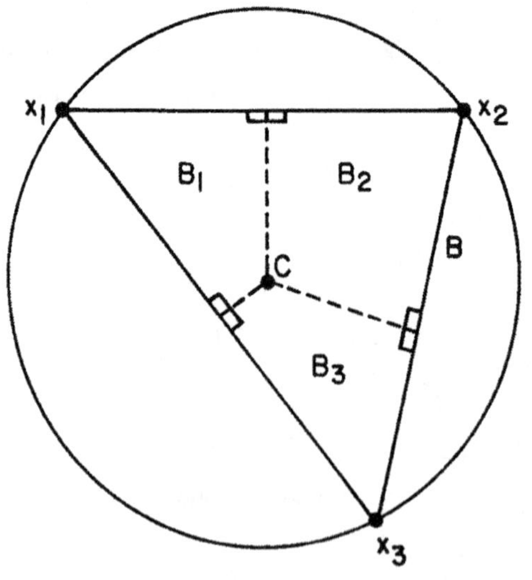

Fig. 1.1

Now

$$x = \lambda_1(x)\, x_1 + \lambda_2(x)\, x_2 + \lambda_3(x)\, x_3 \tag{1.4}$$

and

$$\ell(x) = \lambda_1 f(x_1) + \lambda_2 f(x_2) + \lambda_3 f(x_3),$$

where $\lambda_i \geq 0$, $i = 1,2,3$ and $\lambda_1 + \lambda_2 + \lambda_3 = 1$.
Thus

$$\begin{aligned} |f(x) - \ell(x)| &\leq \sum_{i=1}^{3} \lambda_i\, |f(x) - f(x_i)| \\ &\leq \sum_{i=1}^{3} \lambda_i\, |x - x_i| = : g(x). \end{aligned}$$

We need to show that $g(x) \leq 1$, $x \in B$. Put $r_i = |x - x_i|$, $i = 1,2,3$.

$$g(x) = \lambda_1 r_1 + \lambda_2 r_2 + \lambda_3 r_3 \leq (\lambda_1 r_1^2 + \lambda_2 r_2^2 + \lambda_3 r_3^2)^{1/2}$$

by Schwarz's inequality. But

$$\begin{aligned} (\sum_{i=1}^{3} \lambda_i r_i^2)^{1/2} &= (\sum_{i=1}^{3} \lambda_i\, [(u-u_i)^2 + (v-v_i)^2])^{1/2} \\ &= (\sum_{i=1}^{3} \lambda_i\, [(u^2 - 2u_i u + u_i^2) + (v^2 - 2v_i v + v_i^2)])^{1/2} \\ &= (1 - (u^2 + v^2))^{1/2} \leq 1\,, \end{aligned}$$

where we have used (1.4) and the fact that $u_i^2 + v_i^2 = 1$, $i = 1,2,3$.

It is clear that this result can be generalized in a straightforward fashion to a simplex, B, in $\mathbb{R}^{n-1}$ where sampling takes place at the n vertices.

If B is an equilateral triangle the optimal sampling points x_1^*, x_2^*, x_3^* are easily determined. The accompanying figure shows their location. The associated intrinsic error is now 1/2 (instead of 1 when sampling takes place at the vertices).

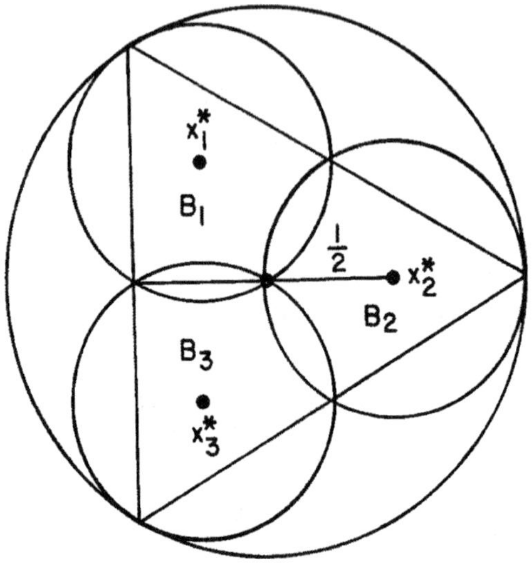

Fig. 1.2

2. General Theory of Optimal Recovery

We now present a broad setting for the optimal recovery problem and a sketch of a general theory. The formulation is taken from M-R.

Let X be a linear space and Y and Z be normed linear spaces. K is a subset of X, U a linear operator from X into Z (the *feature* operator) and I a linear operator from X into Y (the *information* operator). Let $\varepsilon \geq 0$ be given, and assume that we know for each $x \in K$ some $y \in Y$ satisfying $\| Ix - y \| \leq \varepsilon$. From this (possibly erroneous) information we wish to determine the best possible estimate of Ux. To this end, if S denotes the unit ball in Y, let α be any function with domain $IK + \varepsilon S$ and range in Z. We call such a function an *algorithm*. This model is represented schematically in the accompanying diagram.

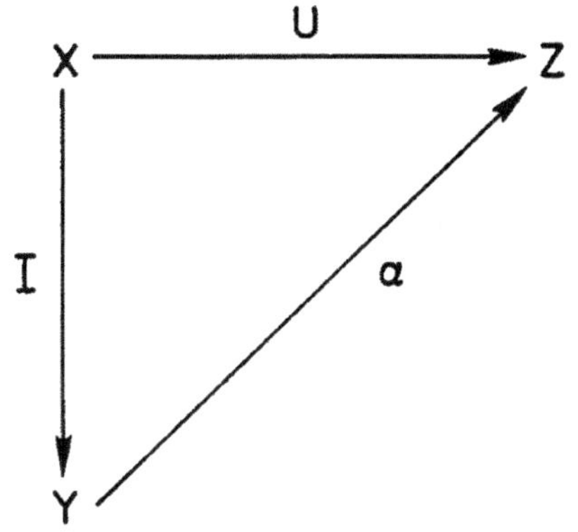

Fig. 2.1

Each algorithm, α, produces an *error*

$$E_\alpha(K, \varepsilon) = \sup \{ \| Ux - \alpha y \| : \ x \in K, \ \| Ix - y \| \le \varepsilon \} ,$$

and

$$E(K, \varepsilon) = \inf_\alpha E_\alpha$$

is called the *intrinsic error* in the problem. If

$$E_{\alpha^*}(K, \varepsilon) = E(K, \varepsilon)$$

then α^* is an *optimal algorithm* and is said to effect the *optimal recovery* of Ux.

Thus in an explicit problem we need to specify X, Y, Z, K, U, I, ε. For example in Problem 1.2 (optimal approximation) we take $X = L^\infty(B)$, K the subset of contractions in X, $Y = \mathbb{R}^n$ (or $\mathbb{C}^n$), Z = X , U is the identity operator, $If = (f(x_1),...,f(x_n))$ and $\varepsilon = 0$.

As we saw in the examples of Section 1, apart from being interesting in its own right, a lower bound to the intrinsic error can be a useful tool in determining an optimal algorithm. Such a lower bound is provided by the following simple result. Recall that a set K is *balanced* if $x \in K$ implies that $-x \in K$.

Theorem 2.1. If K is a balanced convex subset of X then

$$e(K, \varepsilon) := \sup \{ \|Ux\| : x \in K, \|Ix\| \leq \varepsilon \} \leq E(K, \varepsilon) . \tag{2.1}$$

Proof. Suppose $x \in K$ and $\|Ix\| \leq \varepsilon$. For any algorithm α we have

$$\|Ux - \alpha 0\| \leq E_\alpha$$

and

$$\|U(-x) - \alpha 0\| = \|Ux + \alpha 0\| \leq E_\alpha .$$

Therefore

$$\|2Ux\| = \|Ux - \alpha 0 + Ux + \alpha 0\| \leq 2E_\alpha ,$$

hence

$$\|Ux\| \leq E_\alpha ,$$

and (2.1) follows.

Note: (2.1) holds if U,I are nonlinear and satisfy $U(-x) = -Ux$ and $\|Ix\| = \|I(-x)\|$.

As a complement to (2.1) we state

Theorem 2.2 (M-R). If K is a balanced convex subset of X then

$$E(K, \varepsilon) \leq 2\, e(K, \varepsilon).$$

Equality does not always hold in (2.1), but in the following case it does. Let $K_y = \{x \in K : Ix = y\}$.

Theorem 2.3. If G is a function from IX into X such that $x - G\, Ix \in K_0$, all $x \in K$, and K is a balanced convex subset of X, then $e(k, 0) = E(K,0)$ and $\alpha = UG$ is an optimal algorithm.

Proof. When $\alpha = UG$ we have

$$\begin{aligned} E_\alpha(K,0) &= \sup_{x\in K} \|Ux - UGIx\| \\ &= \sup_{x\in K} \|U(x - GIx)\| \\ &\le \sup_{x\in K_0} \|Ux\| = e(K,0) . \end{aligned}$$

Example 2.1. Consider the optimal recovery problem specified as follows: Suppose $n \ge 1$,

$$X = \{f \in L^2[0,1] : f \in C^{n-1}[0,1], f^{(n)} \in L^2[0,1]\} ,$$

$Z = X$, U is the identity operator and $K = \{f \in X : \|f^{(n)}\|_2 \le 1\}$. Moreover given $0 \le t_1 < t_2 <,\ldots,< t_k \le 1$, $k \ge n$, put $Y = \mathbb{R}^k$, $If = (f(t_1),\ldots,f(t_k))$ and $\varepsilon = 0$.

Recall that for each $f \in X$ there is a natural spline of order $2n$ with knots at $t : (t_1,\ldots,t_k)$, S_f, which satisfies

$$S_f(t_i) = f(t_i) , i = 1,\ldots,k$$

and, moreover, $S_f \in X$ and

$$\|f^{(n)} - S_f^{(n)}\| \le \|f^{(n)}\| . \tag{2.2}$$

Define G by

$$If \overset{G}{\rightarrow} S_f .$$

Then $I(f - GIf) = If - IS_f = 0$. Also, in view of (2.2), if $f \in K$ then $f - GIf \in K$. Thus the hypotheses of Theorem 2.3 are satisfied and natural spline interpolation is optimal. Generalizations of this result are given later.

Next we examine the relationship between the optimal recovery problem and the notion of a Chebyshev center. Recall that if M is a bounded subset of Z then $z_0 \in Z$ is called a Chebyshev center for M if

$$\sup_{a\in M} \|z_0 - a\| = \inf_{z\in Z} \sup_{a\in M} \|z - a\| =: r(M).$$

r(M) is called the Chebyshev radius of M. The "hypercircle" H(y) is defined by

$$H(y) = \{Ux : x \in K_y\} .$$

We can now state

Theorem 2.4 (M-R). Suppose that for every $y \in \mathbb{K}$, $H(y)$ has a Chebyshev center, $c(y) \in Z$. Then $\alpha^* : y \to c(y)$ is an optimal algorithm and

$$E(K,0) = \sup_{y \in \mathbb{K}} r(H(y)).$$

Example 2.2. $X = Z = C[0,1]$ and K is the set of contractions on $[0,1]$. $Y = \mathbb{R}^2$, U is the identity operator, $If = (f(0), f(1))$ and $\varepsilon = 0$. We saw in Section 1 that linear interpolation was an optimal algorithm for this problem. We shall now show that Theorem 2.4 produces another optimal algorithm which, however, is nonlinear in the data. Namely put $\mu(t) = \min\{y_0 + t, y_1 + (1-t)\}$, $\lambda(t) = \max\{y_0 - t, y_1 - (1-t)\}$, where $y_0 = f(0)$, $y_1 = f(1)$. Then for each $y : (y_0, y_1)$ in $\mathbb{K}$, $H(y) = K_y$ has a Chebyshev center $c(y) \in C[0,1]$ given by

$$c(y)(t) = \frac{1}{2}(\mu(t) + \lambda(t)), \quad 0 \le t \le 1.$$

Since

$$r(H(y)) = \frac{1}{2} - \frac{|y_1 - y_0|}{2},$$

we obtain $E(K,0) = 1/2$. $y \to c(y)$ is an optimal algorithm. Examination of Fig. 2.2 confirms these conclusions.

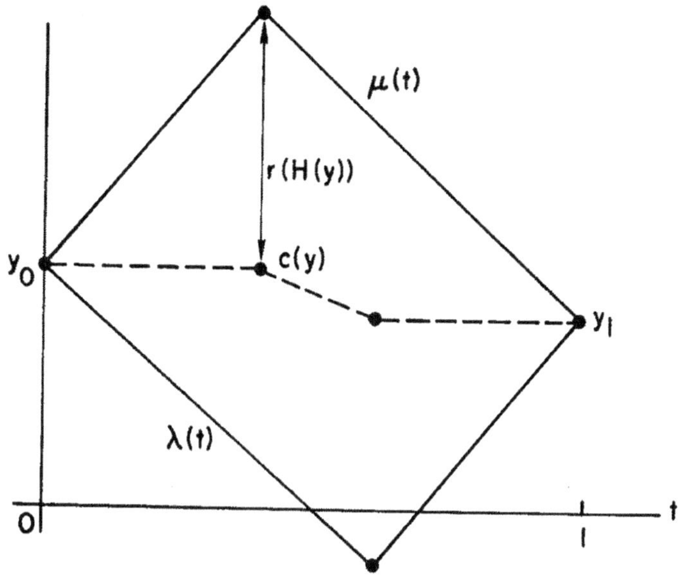

Fig. 2.2

Optimal algorithms which are linear need not always exist. But in some important cases we can guarantee their existence. For example, suppose X is a linear space over the reals and Z = ℝ. Then U is a linear functional. Suppose K is balanced and convex and let Y^* denote the conjugate space of Y. Recall that $A \subset Y$ is called an *absorbing set* in Y if for every $y \in Y$ there exists $\lambda > 0$ such that $\lambda y \in A$. We now state some results (details in M-R. See also Scharlach [50].) in this context.

Theorem 2.5. If $\varepsilon > 0$, or $\varepsilon = 0$ and IK is an absorbing set in Y then

$$e(K,\varepsilon) = E(K,\varepsilon) = \inf_{L \in Y^*} \sup \{ |Ux - Ly| : x \in K , \|Ix - y\| \le \varepsilon \} .$$

Theorem 2.6. If there exists $L \in Y^*$ such that

$$\sup_{x \in K} |Ux - LIx| < \infty$$

and $\varepsilon > 0$, or $\varepsilon = 0$ and IK is a neighborhood of the origin in Y, then an optimal algorithm which is in Y^* exists.

$x_0 \in K$ is a *worst function* if

$$\|Ux_0\| = \sup\{ \|Ux\| : x \in K , \|Ix\| \le \varepsilon\} .$$

We wish to describe the relationship between the following statements:

(a) x_0 is a worst function.

(b) $L_0 \in Y^*$ is an optimal algorithm.

(c) $\varepsilon \|L_0\| = L_0 I x_0$ and

$$\max_{x \in K} |Ux - L_0 Ix| = Ux_0 - L_0 Ix_0.$$

Theorem 2.7. If (a) holds then (b) and (c) are equivalent. If (b) holds then (a) and (c) are equivalent.

Similar results hold for complex-valued functions and linear functionals (See M-R).

Example 2.3. Let B denote the interval $[-1,1]$. $X = L^\infty(B)$, $n \ge 1$, $K = \{f \in C^{n-1}(B) : |f^{(n-1)}(u) - f^{(n-1)}(v)| \le |u-v|;\ u, v \in B\}$, $Y = \mathbb{R}^n$ (equipped with the max norm), $Z = \mathbb{R}$. $-1 \le t_1 < \ldots < t_{n+r} \le 1$, $If = (f(t_1),\ldots,f(t_{n+r}))$, $Uf = f(\tau)$, $\tau \in B$ and $\varepsilon \ge 0$.

We wish to show that when $r = 0$ polynomial interpolation is optimal in this problem. Let $\mathscr{P}_k$ denote the set of polynomials of degree at most k. Let ℓ_i, $i = 1,\ldots,n$ be the fundamental polynomials for interpolation at $t_1,\ldots,t_n$. That is, $\ell_i \in \mathscr{P}_{n-1}$ and $\ell_i(t_j) = \delta_{ij}$, $i,j = 1,\ldots,n$. Put

$$p(t) = \varepsilon \sum_{i=1}^{n} (\operatorname{sgn} \ell_i(\tau))\, \ell_i(t)$$

and

$$Q(t) = \frac{(t-t_1)\ldots(t-t_n)}{n!} .$$

Note that Q and $g(t) = Q(t) \operatorname{sgn} Q(\tau) + p(t)$ are both in K. Also

$$\|Ig\| = \max_i |g(t_i)| = \varepsilon.$$

Therefore

$$(2.3) \qquad e(K, \varepsilon) \geq |g(\tau)| = |Q(\tau)| + \varepsilon \sum_{i=1}^{n} |\ell_i(\tau)| .$$

Suppose y: $(y_1,...,y_n) \in Y$. Put

$$v(t) = \sum_{i=1}^{n} y_i \, \ell_i(t)$$

and

$$u(t) = \sum_{i=1}^{n} f(t_i) \, \ell_i(t) .$$

Consider the algorithm α : $y \to v(\tau)$. Suppose $|f(t_i)-y_i| \leq \varepsilon$, $i = 1,...,n$. Now $|f(\tau)-v(\tau)| \leq |f(\tau)-u(\tau)| + |u(\tau)-v(\tau)|$. But for $f \in K$, $|f(\tau) - u(\tau)| \leq |Q(\tau)|$, and since

$$|u(\tau) - v(\tau)| \leq \sum_{i=1}^{n} |f(t_i)-y_i| \, |\ell_i(\tau)| \leq \varepsilon \sum_{i=1}^{n} |\ell_i(\tau)|$$

we conclude that $E_\alpha(K,\varepsilon) \leq |g(\tau)|$. Thus $E_\alpha(K,\varepsilon) = E(K,\varepsilon) = e(K,\varepsilon) = |g(\tau)|$.

Observe that $x_0 = g$ is a worst function and $L_0 = \alpha$ is an optimal algorithm.

$$\|L_0\| = \max_{\|y\| \leq 1} |v(\tau)| = \sum_{i=1}^{n} |\ell_i(\tau)|.$$

$Ig = \varepsilon(\mathrm{sgn}\, \ell_i(\tau),...,\mathrm{sgn}\, \ell_n(\tau))$ and so $\varepsilon \|L_0\| = L_0 I x_0$. Thus Theorems 2.5 - 2.7 are easily verified.

The case $r \geq 1$ is analyzed in Micchelli [32] where the following result is proved.

Theorem 2.8. There exists an integer k, $0 \leq k \leq r$, a subset $\{s_1,...,s_{n+k}\}$ of $\{t_1,...,t_{n+r}\}$ and a subset $\{x_1,...,x_k\}$ of $(-1,1)$ with the property that for any data $y_1,...,y_{n+k}$ there is a unique spline,

$$S(t) = \sum_{j=0}^{n-1} a_j \, t^j + \sum_{j=1}^{k} c_j(t - x_j)_+^{n-1} ,$$

satisfying $S(s_i) = y_i$, $i = 1,...,n + k$, and α : $(y_1,...,y_{n+k}) \to S(\tau)$ is an optimal algorithm in estimating $f(\tau)$.

Some information about k in terms of $t_1,...,t_{n+r}$ and ε is known. For instance, $k = r$ when $\varepsilon = 0$ (cf. Micchelli, Rivlin and Winograd [36]).

In the case $n = 1$ the optimal algorithms can be fully described. If we choose $x_i = (t_i + t_{i+1})/2$, $i = 1,\dots,r$ and $x_0 = -1$, $x_{r+1} = 1$, then if $x_{i-1} < \tau < x_i$, $i = 1,\dots,r+1$ (or $\tau = x_0, x_{r+1}$) $\alpha : (y_1,\dots,y_{r+1}) \to y_i$ is the unique linear optimal algorithm. Furthermore, if $\tau = x_i$, $i = 1,\dots,r$, the linear optimal algorithms are exactly those which produce any convex combination of y_i and y_{i+1}.

Finally, we present an example in which we describe the optimal recovery of a point on a curve in the plane from knowledge of a finite set of points on the curve and the set of curves under consideration. We shall see that there is a linear optimal algorithm for this problem, and that the problem may be reduced to its one dimensional components.

Example 2.4. Suppose $n \ge 1$ and $0 \le t \le 1$. Let $x(t) = (x_1(t), x_2(t))$, $0 \le t \le 1$. Put $K = \{x(t) : |x^{(n)}(t)|_2 \le 1, 0 \le t \le 1\}$ where $|\cdot|_2$ is the euclidean norm in the plane. Suppose $0 < t_1 < t_2 < \dots < t_{n+r} < 1$ and $Ix = (x(t_1),\dots,x(t_{n+r}))$. Given $0 \le \tau \le 1$ we seek to recover $x(\tau)$ from Ix ($\varepsilon = 0$) for $x \in K$. (2.1) yields

$$e = e(K,0) = \sup\{|x(\tau)|_2 : x \in K, x(t_i) = 0, i = 1,\dots,n+r\}.$$

In the previous example we mentioned that the scalar version of this problem was solved in Micchelli, Rivlin and Winograd [36], i.e., they determined

$$\tilde{e} = \sup\{|f(\tau)| : \|f^{(n)}\|_\infty \le 1, f(t_i) = 0, i = 1,\dots,n+r\},$$

and a linear optimal algorithm for this problem. Clearly, if we let $x(t) = f(t)v$ where $f(t_i) = 0$, $i = 1,\dots,n+r$, $\|f^{(n)}\|_\infty \le 1$ and $|v|_2 = 1$ we see that $\tilde{e} \le e$. To prove the reverse inequality choose any v such that $|v|_2 = 1$ and set $f(t) = (v, x(t))$ where $x \in K$ and $x(t_i) = 0$, $i = 1,\dots,n+r$. Then $|f^{(n)}(t)| \le 1$, $0 \le t \le 1$, by Schwarz's inequality, and $f(t_i) = 0$, $i = 1,\dots,n+r$. Hence $|(v, x(\tau))| \le \tilde{e}$ and so $\sup_v |(v,x(\tau)| \le \tilde{e}$, which implies that $e \le \tilde{e}$, and, therefore, $e = \tilde{e}$.

Let $a_1,\dots,a_{n+r}$ be the coefficients in the linear optimal algorithm α, for the scalar version, so that for any f satisfying $\|f^{(n)}\|_\infty \le 1$ we have

$$|f(\tau) - \sum_{i=1}^{n+r} a_i f(t_i)| \le \tilde{e}. \tag{2.3}$$

Then for any v satisfying $|v|_2 = 1$ and $x \in K$ we put $f(t) = (v, x(t))$ in (2.3) and obtain

$$|(v, x(\tau) - \sum_{i=1}^{n+r} a_i x(t_i))| \le \tilde{e} = e.$$

Taking the supremum over v and then over x now yields

$$E_\alpha(K,0) = \sup_{x \in K} \left| x(\tau) = \sum_{i=1}^{n+r} a_i \, x(t_i) \right| \leq e \, .$$

Thus $e = \tilde{e}$ is the intrinsic error and the coefficients of α provide a linear optimal algorithm in our problem.

3. Optimal Recovery in Hilbert Spaces

We now consider the optimal recovery problem in the case that one, or more, of the spaces X, Y, Z is a Hilbert space. We shall emphasize the existence of linear optimal algorithms in that setting. But we begin with the negative observation that when $\dim Z > 1$ there may not be any linear optimal algorithms, as the following example shows.

Example 3.1. (Melkman and Micchelli [29]). Let $X = \mathbb{R}^2$ with $\|x\|_X^2 = x_1^2 + x_2^2$. Take $K = \{x \in X : \|x\| \leq 1\}$. Let $Ux = x$ map X into $Z = \mathbb{R}^2$, normed by $\|x\|_Z^4 = \lambda_1 x_1^4 + \lambda_2 x_2^4$, $\lambda_1 > \lambda_2 > 0$. Suppose $Ix = x_1$, so that $Y = \mathbb{R}$. Then for $\varepsilon \geq 0$, but sufficiently small, $E(K,\varepsilon) = e(K,\varepsilon)$ and $E^4(K,\varepsilon) = \lambda_2$. However, while the intrinsic error is attained for a nonlinear algorithm no linear algorithm can do as well.

Despite this simple example there are quite general settings in which a linear optimal algorithm exists. One such was described in Theorem 2.6. However, here we wish first to consider Theorem 2.3 in the case that X is a Hilbert space and K is the unit ball in X. If $N(I) = \{x \in X : Ix = 0\}$ is closed and P is the orthogonal projection of X on the subspace N(I), that is, $\|x - Px\| = \min\{\|x-u\| : u \in N(I)\}$, then since $\|Px\| \leq \|x\|$ we have $Px \in K_0$ for $x \in K$. Thus $G : Ix \to x - Px$ is a linear operator from IX into X which satisfies the hypotheses of Theorem 2.3 and $\alpha = UG$ is an optimal algorithm in the case of exact information ($\varepsilon = 0$). It is important to note that this result holds for any normed linear space Z.

This result has wide application. We shall give three applications shortly. Since specification of U plays no role in the result we make no reference to it in these examples.

First a general remark about identifying $Qx = x - Px$. When Y is also a Hilbert space and I is a bounded linear operator from X to Y whose adjoint, I^*, is easily identifiable then Q can be determined as the orthogonal projection of X onto $\mathcal{R}(I^*)$. Therefore when $\dim Y = n$ and $Ix = ((x_1,x),\ldots,(x_n,x))$ for linearly independent $x_1,\ldots,x_n \in X$ then

$$I^* y = \sum_{i=1}^{n} y_i x_i$$

and the orthogonal projection Q is given by

$$Q = \sum_{i,j} g_{ij}^{-1} \, x_i \otimes x_j$$

where (g_{ij}) is the Gramian matrix (i.e., $g_{ij} = (x_i,x_j)$) and $(x_i \otimes x_j)\, x = x_i\, (x_j,x)$. Consequently the corresponding optimal algorithm is

$$\alpha : (y_1,\dots,y_n) \to \sum_{j=1}^{n} y_j \,(\sum_{i=1}^{n} g_{ij}^{-1}\, Ux_i)$$

These formulas are particularly useful when X is a reproducing kernel Hilbert function space and $If = (f(t_1),\dots,f(t_n))$. For then $If = ((K(t_1,t), f(t)),\dots,(K(t_n,t), f(t)))$. The Gramian matrix is now $(K(t_i, t_j))$ and

$$(Qf)(t) = \sum_{i=1}^{n} a_i\, K(t,t_i)\ ,$$

where the a_i are determined by the condition $(Qf)(t_i) = f(t_i)$.

Our next example is useful in mathematical studies of computer assisted tomography.

Example 3.2. $\Delta = \{(x,y) : x^2 + y^2 = 1\}$ and $X = L^2(\Delta)$. For every $\theta \in (-\pi,\pi]$ and $\rho \in [-1,1]$ let

$$(I_\theta f)(\rho) = \int_{x \cos\theta + y \sin\theta = \rho} f(x,y)dxdy$$

$$= \int_{-\sqrt{1-\rho^2}}^{\sqrt{1-\rho^2}} f(\rho \cos\theta - s \sin\theta,\ \rho \sin\theta + s \cos\theta)ds.$$

Then $I_\theta : L^2(\Delta) \to L^2[-1,1]$, and its adjoint is given by

$$(I^*_\theta g)(x,y) = g(x \cos\theta + y \sin\theta)\ , \qquad g \in L^2[-1,1],$$

because

$$\int_{-1}^{1} g(\rho)(I_\theta f)(\rho)d\rho = \iint_{x^2+y^2 \le 1} f(x,y)g(x \cos\theta + y \sin\theta)dxdy.$$

In the terminology of Logan and Shepp [24] $I^*_\theta g$ is called a *ridge function* (F. John calls it a *plane wave*). We define $I = (I_{\theta_1},\dots,I_{\theta_n})$ and

$Y = L^2[-1,1] \times \ldots \times L^2[-1,1]$. Then

$$I^*(g_1,\ldots,g_n)(x,y) = \sum_{i=1}^{n} g_i(x \cos \theta_i + y \sin \theta_i)$$

and so

$$(Qf)(x,y) = \sum_{i=1}^{n} g_i\,(x \cos \theta_i + y \sin \theta_i)$$

where $g_1,\ldots,g_n \in L^2[-1,1]$ are chosen so that $IQf = If$. The explicit construction of Q for equally spaced angles is given in Logan and Shepp [24]. When $n = 1$ it is easy to see that

$$(Qf)(x,y) = \frac{1}{2(1-\rho^2)^{1/2}} (I_{\theta_1} f)(x \cos \theta_1 + y \sin \theta_1).$$

The importance of this example in tomography is discussed in Logan and Shepp [24].

Example 3.3. $X = W_2^m(\mathbb{R}^n)$, $m > n/2$. The norm in the Hilbert space X is given by

$$\| f \|^2 = \sum_{|\alpha|=m} \int_{\mathbb{R}^n} \binom{m}{\alpha} \left| \frac{\partial^\alpha f}{\partial x^\alpha} \right|^2 dx.$$

Here $\alpha = (\alpha_1,\ldots,\alpha_n)$, $|\alpha| = \alpha_1 + \ldots + \alpha_n$, and

$$\binom{m}{\alpha} = \frac{m!}{\alpha_1!\ldots\alpha_n!} .$$

Let $If = (f(x_1),\ldots,f(x_N))$ for given $x_i \in \mathbb{R}^n$. We further require that if $r \in \mathscr{P}_{m-1}(\mathbb{R}^n)$ (polynomials of total degree $\leq m-1$) and $r(x_i) = 0$, $i = 1,\ldots,N$ then $r = 0$. Then

$$(Qf)(x) = p(x) + \sum_{i=1}^{N} a_i(f)\, \varphi\,(x - x_i) ,$$

where $p \in \mathscr{P}_{m-1}(\mathbb{R}^n)$,

$$\varphi(x) = \begin{cases} \|x\|^{2m-n} \log \|x\| , & n \text{ even} \\ \|x\|^{2m-n} , & n \text{ odd} , \end{cases}$$

and $\|x\|$ is the euclidean norm of x. In addition

$$(Qf)(x_i) = f(x_i) , \; i = 1,\ldots,N$$

and

$$\sum_{i=1}^{N} a_i q(x_i) = 0 , \qquad q \in \mathscr{P}_{m-1}(\mathbb{R}^n) .$$

Details in Duchon [10] and Meinguet [26]. When n = 1, Qf is a natural spline of order 2m and we recover Example 2.1.

Example 3.4. (Melkman [27]) $X = \{f \in L^2(\mathbb{R}) : \hat{f}(\omega) = 0, \ |\omega| > \sigma\}$ where $\hat{f}$ denotes the Fourier transform of f. Thus X is the space of band-limited functions. Let $Y = \mathbb{R}^n$ and $If = (f(t_1),\ldots,f(t_n))$, then

$$I^*(\alpha_1,\ldots,\alpha_n)(t) = \sum_{i=1}^{n} \alpha_i \frac{\sin 2\pi\sigma(t-t_i)}{\pi(t-t_i)}$$

because

$$\sum_{i=1}^{n} \alpha_i f(t_i) = \left(\sum_{i=1}^{n} \alpha_i \frac{\sin 2\pi\sigma(t-t_i)}{\pi(t-t_i)} , f(t)\right)_X .$$

Hence Qf is interpolation by a linear combination of the functions

$$\frac{\sin 2\pi\sigma(t-t_i)}{\pi(t-t_i)} , \qquad i = 1,\ldots,n,$$

a well-known procedure.

Next we turn to a result on the existence of linear optimal algorithms in the presence of inaccurate information ($\varepsilon > 0$).

Theorem 3.1 (Melkman and Micchelli [29] Suppose X, Y, Z are Hilbert spaces, U is a linear operator and I is a bounded linear operator. Then, for $\varepsilon \geq 0$, $E(K,\varepsilon) = e(K,\varepsilon)$ and there is a linear optimal algorithm.

The proof of this theorem provides a construction of the optimal algorithm. The procedure is the following. For $\varepsilon > 0$ and $0 \leq \mu \leq 1$ we define $x(\mu) = J_\mu y$ to be a solution to the minimum problem

$$\min_{x\in X} \{(1-\mu)\, \varepsilon^{-2} \, \|Ix - y\|_Y^2 + \mu \, \|x\|_X^2\} .$$

The function $\Psi(\mu) = \max \{ \|Ux\| : \mu \|x\|^2 + (1-\mu)\, \varepsilon^{-2} \|Ix\|^2 \leq 1\}$ is convex on $[0,1]$ and $\min \{\Psi(\mu) : 0 \leq \mu \leq 1\} = e(K,\varepsilon)$. Moreover, if $\Psi(\lambda) = e(K,\varepsilon)$ then $\alpha(y) = UJ_\lambda y$ is a linear optimal algorithm.

Example 3.5. Let $X = Z = \tilde{L}^2[0,1]$, one-periodic L^2 functions on $[0,1]$, and $Uf = f$. Also choose $If = (f_{-k},...,f_k)$ where

$$f_j = \int_0^1 f(t)e^{2\pi ijt}dt ,$$

$Y = \mathbb{C}^{2k+1}$, with the euclidean norm and

$$K = \{f \in X : \int_0^1 (f'(t))^2 dt \leq 1\} .$$

Since

$$\int_0^1 (f'(t))^2 dt = 4\pi^2 \sum_{j=-\infty}^{\infty} j^2 |f_j|^2$$

we obtain

$$\mu \int_0^1 (f')^2 + (1-\mu)\, \varepsilon^{-2} \sum_{|j| \leq k} |f_j|^2 \geq \mu(4\pi^2) \sum_{|j| \geq k+1} j^2 |f_j|^2 + (1-\mu)\, \varepsilon^{-2} \sum_{|j| \leq k} |f_j|^2$$
$$\geq \min\left((1-\mu)\, \varepsilon^{-2}, 4\pi^2\mu\, (k+1)^2\right) \sum_{j=-\infty}^{\infty} |f_j|^2 ,$$

with equality possible. Thus

$$\Psi^2(\mu) = \frac{1}{\min\left((1-\mu)\varepsilon^{-2}, 4\pi^2\mu(k+1)^2\right)} ,$$

has a unique minimum at

$$\lambda = \frac{1}{1 + 4\pi^2\varepsilon^2(k+1)^2} ,$$

and so

$$e^2(K,\varepsilon) = \Psi^2(\lambda) = \varepsilon^2 + \frac{1}{4\pi^2(k+1)^2} ,$$

a bound which is a special case of Lemma 4 in M-R.

Moreover, it is easy to see that $J_\lambda y$ is a trigonometric polynomial of degree k whose coefficients are given by

$$\widehat{(J_\lambda y)}_j = \frac{(k+1)^2}{(k+1)^2 + j^2}\, y_j , \ |j| \leq k .$$

Note that the optimal algorithm is independent of ε .

For $\varepsilon = 0$ our previous remarks tell us that

$$(\hat{\alpha y})_j = \begin{cases} y_j , & |j| \le k \\ 0 , & |j| > k \end{cases}$$

is an optimal algorithm. As was observed in Remark 2 of Section 4 in M-R, this algorithm is also optimal for $\varepsilon > 0$ because

$$E_\alpha^2(K,\varepsilon) = \sup\{ \sum_{|j|\le k} |f_j - y_j|^2 + \sum_{|j|>k} |f_j|^2 : \sum_{|j|\le k} |f_j - y_j|^2 \le \varepsilon^2 ; 4\pi^2 \sum j^2 |f_j|^2 \le 1\}$$

$$\le \varepsilon^2 + \frac{1}{4\pi^2(k+1)^2} = e^2(K,\varepsilon) .$$

It is easy to construct many other optimal algorithms in this case. Other examples of Theorem 3.1 are given in Melkman and Micchelli [29].

Next we sketch the proof of Theorem 3.1. There are two key ingredients in the proof. First we show that for any $\mu \in [0,1]$ there is a linear algorithm with error $\Psi(\mu)$. To this end we introduce the product space $W = X \times Y$ with Hilbert space semi-norm

$$\|w\|_\mu^2 = \mu \|x\|_X^2 + (1-\mu)\, \varepsilon^{-2} \|t\|_Y^2$$

where $w = (x,t)$. Let $\hat{K}$ be the unit ball in W, and set $\hat{U}w = Ux$, $Iw = Ix - t$ and $\hat{\varepsilon} = 0$. The diagrams of the original and "hatted" problems are shown in Figures 3.1a,b respectively.

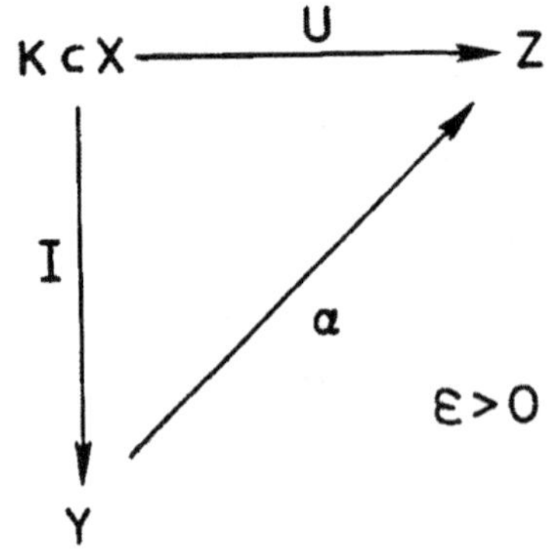

Fig. 3.1a

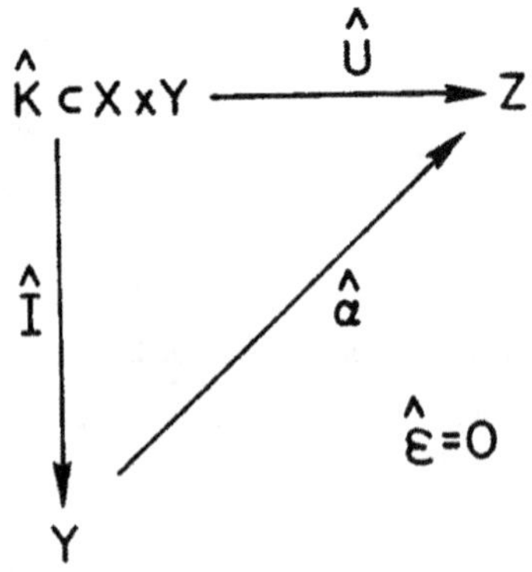

Fig. 3.1b

The linear optimal algorithm for estimating $\hat{U}w$ from exact information, $\hat{I}w = y$ is obtained, using the remarks at the beginning of this section, by solving the minimum problem.

$$\min_{\hat{I}w=y} \|w\|_{\mu}^2 .$$

But this is equivalent to solving the minimum problem

$$\min_{x} \{\mu \|x\|_X^2 + (1-\mu)\, \varepsilon^{-2} \|Ix - y\|^2\} .$$

If we put $\hat{\alpha}y = (x,t)$ where $x = J_\mu y$ and bear in mind that $Ix - t = y$ then we have

$$\begin{aligned} E(\hat{K},0) = e(\hat{K},0) &= \{\sup \|\hat{U}w\|_Z : \hat{I}w = 0 , \|w\|_\mu \leq 1\} \\ &= \{\sup \|Ux\|_Z : \mu \|x\|_X^2 + (1-\mu)\, \varepsilon^{-2} \|Ix\|_Y^2 \leq 1\} \\ &= \Psi(\mu) . \end{aligned}$$

Therefore,

$$\begin{aligned} \Psi(\mu) &= E_\mu(K,\varepsilon) = \sup\{\|\hat{U}w - \hat{U}\hat{\alpha}\hat{I}w\|_Z : \|w_\mu\| \leq 1\} \\ &= \sup\{\|Ux - U J_\mu(Ix - t)\|_Z : \mu \|x\|_X^2 + (1-\mu)\, \varepsilon^{-2} \|t^2\|_Y \leq 1\} \\ &= \sup\{\|Ux - U J_\mu y\|_Z : \mu \|x\|_X^2 + (1-\mu)\, \varepsilon^{-2} \|Ix - y\|_Y^2 \leq 1\} \\ &\geq \sup\{\|Ux - UJ_\mu y\|_Z : \|x\|_X \leq 1 , \|Ix - y\| \leq \varepsilon\} \\ &= E(\alpha_\mu, \varepsilon) . \end{aligned}$$

The second part of the proof is to show that

$$\min_{\mu\epsilon[0,1]} \Psi(\mu) = \sup\{\|Ux\|_Z : \|x\|_X \leq 1 , \|Ix\|_Y \leq \varepsilon\} = e(K,\varepsilon) .$$

This is equivalent to showing that

$$\inf_{a\epsilon B} \|a\|_\infty = \max_{\gamma\epsilon\Delta} \inf_{a\epsilon B} \gamma \bullet a,$$

where $a = (a_1,a_2)$, $\gamma = (\gamma_1,\gamma_2)$, $\Delta = \{\gamma : |\gamma_1| + |\gamma_2| \leq 1\}$, $\|a\|_\infty = \max (a_1,a_2)$ and $B = \{(\|x\|_X^2 , \varepsilon^{-2} \|Ix\|_Y^2) : \|Ux\|_Z^2 = 1\}$.

Although B is not a convex set the required equality follows because B shows a *convex profile* to the origin. That is, whenever a supporting line touches B at two points the line segment joining these two points is in B. This property of B follows easily from the fact that all the norms used to define B are quadratic. See Melkman and Micchelli [29] for further details.

4. Stochastic Information

In our previous discussion of optimal recovery from inaccurate information our point of view was that the inaccurate data $y \in Y$ had the form $y = Ix + e$, where Ix is the exact data corresponding to some $x \in K$ and e is in the ε-ball about the origin in Y. From this point of view every e in this ε-ball is equally likely. We now consider the possibility that e is a random variable. Now our error criterion for an algorithm α is the worst expected value of the square of the discrepancy, i.e.,

$$\sup_{x \in K} \mathscr{E}\left(\| Ux - \alpha(Ix + e) \|^2 \right),$$

and our objective is to minimize this error over all α. In this generality we are unable to solve even a special case of our problem. We must take a significant departure from our previous approach and assume that only linear algorithms, α, be considered. For simplicity we also assume that $Y = \mathbb{R}^n$ Furthermore, in order to simplify the expected mean square error we assume that Z as well as X is a Hilbert space. If in addition $e = (e_1, \ldots, e_n)$ has mean zero, $\mathscr{E}(e_i) = 0$, $i = 1, \ldots, n$, known covariance $\mathscr{E}(e_i e_j) = c_{ij}$ and

$$\alpha(r_1, \ldots, r_n) = \sum_{i=1}^{n} z_i r_i \; ; \qquad r_i \in \mathbb{R}, \; z_i \in Z,$$

we obtain

$$\mathscr{E}(\| Ux - \alpha\,(Ix + e) \|^2) = \mathscr{E}(\| Ux - \alpha Ix \|^2 - 2(Ux, \alpha e) + \| \alpha e \|^2)$$

$$= \| Ux - \alpha Ix \|^2 + \sum_{i,j} c_{ij}\,(z_i, z_j)$$

$$= \| Ux - \sum_{i=1}^{n} z_i (x_i, x) \|^2 + \sum_{i,j} c_{ij} (z_i, z_j),$$

where $Ix = ((x_1, x), \ldots, (x_n, x))$.

For white noise, i.e., $c_{ij} = \sigma^2\delta_{ij}$, this problem is formulated in Micchelli and Wahba [39], and in a special case is related to a minimax regression problem studied in Laüter [21]. In general the best choice of $z_1,\ldots,z_n$ has not been determined even when K is the unit ball in a Hilbert space X. However, when $Z = \mathbb{R}$ this problem was solved in Speckman [51]. We prove the following theorem by following the method used in Micchelli and Wahba [39] for the white noise case. We define $\langle u,v\rangle_c = \Sigma\, c_{ij}^{-1}\, u_i\, v_j$, $u = (u_1,\ldots,u_n)$, $v = (v_1,\ldots,v_n)$.

Theorem 4.1. Let X be a Hilbert space. Suppose $x_c(y) \in X$ minimizes

$$\|x\|_X^2 + \langle Ix - y, Ix - y\rangle_c$$

then

$$(Ux_c)(y) = \sum_{i=1}^{n} a_i^0 y_i\ ,\ a_i^0 \in \mathbb{R},\ i = 1,\ldots,n$$

and $(a_1^0,\ldots,a_n^0)$ minimizes

$$\sup_{\|x\|_X \le 1} |Ux - \sum_{i=1}^{n} a_i(x_i, x)| + \sum_{i,j} c_{ij}a_ia_j$$

over all $(a_1,\ldots,a_n) \in \mathbb{R}^n$.

Proof. As in the proof of Theorem 3.1 we consider the space $W = X\times Y$ with semi-norm given by $\|w\|^2 = \|x\|^2 + \langle t,t\rangle_c$, where $w = (x,t)$. As before we set $\hat{U}w = Ux$ and $\hat{I}w = Ix - t$. Then it is not difficult to see that

$$\sup_{\|w\|^2 \le 1} |\hat{U}w - \sum_{i=1}^{n} a_i\hat{I}_iw|^2 = \sup_{\|x\| \le 1} |Ux - \sum_{i=1}^{n} a_iI_ix|^2 + \sup_{\langle t,t\rangle_c \le 1} |\sum_{i=1}^{n} a_it_i|^2$$

$$= \sup_{\|x\| \le 1} |Ux - \sum_{i=1}^{n} a_i(x_i,x)|^2 + \sum_{i,j} c_{ij}a_ia_j\ .$$

Thus by our previous remarks the minimum is obtained from $\hat{U}w_0$ where $w_0 = (x_c(y),t)$, $Ix_c(y) - y = t$ and w_0 minimizes $\{\|w\|^2 : \hat{I}w = y\}$. "Dehatting" now proves the theorem.

Example 4.1. $X = L^2[0,1]$,

$$K = \{f \in X : \int_0^1 (f^{(k)}(t))^2 dt \le 1\}\ ,$$

$Y = \mathbb{R}^n$, $Z = \mathbb{R}$, If $= (f(t_1),...,f(t_n))$, $y_i = f(t_i) + e_i$, $\mathscr{E}(e_i e_j) = c_{ij} = \sigma^2 \delta_{ij}$, $\mathscr{E}(e_i) = 0$, $i, j = 1,...,n$. $\tau \in [0,1]$ and $Uf = f(\tau)$. The linear optimal estimator is $f_\sigma(\tau)$ where f_σ is the "smoothing spline" minimizing

$$\| f^{(k)} \|^2 + \sigma^{-2} \sum_{i=1}^{n} (f(t_i) - y_i)^2 .$$

Note that the smoothing parameter, σ^{-2}, is independent of τ. The corresponding problem with "conventional" inaccurate data as described in Theorem 3.1 (with $Y = \ell_n^2$) which also is solved by a smoothing spline, whose smoothing parameter, γ, in the expression

$$\| f^{(k)} \|^2 + \gamma \sum_{i=1}^{n} (f(t_i) - y_i)^2 ,$$

however, depends on ε and τ.

5. Restricted Estimation

In this section we elaborate on our discussion in Section 5 of M-R. Note that in Example 4.1 optimal algorithms come from a spline space of dimension equal to the number of data points. If this number is large these methods may be computationally expensive. For this reason it seems appropriate to restrict the form of the algorithms. To this end we suppose that X and Z are Hilbert spaces, U is a bounded linear operator, $Y = \mathbb{R}^n$ and $Ix = ((v_1,x),...,(v_n,x))$ for $v_1,...,v_n \in X$. Let $\mathscr{V} = \text{span}\,(v_1,...,v_n)$. In addition, we choose an m-dimensional subset, $\mathscr{U}$, of Z spanned by $u_1,...,u_m$. Let A denote any mapping from $\mathbb{R}^n$ into $\mathbb{R}^m$ and suppose $M : \mathbb{R}^m \to Z$ is given by $M(b_1,...,b_m) = \Sigma\, b_i u_i$. We now propose to estimate Ux, for $x \in K = \{x \in X : \|x\| \le 1\}$, by an algorithm of the restricted form M A Ix. Thus our diagram now is shown in Fig. 5.1.

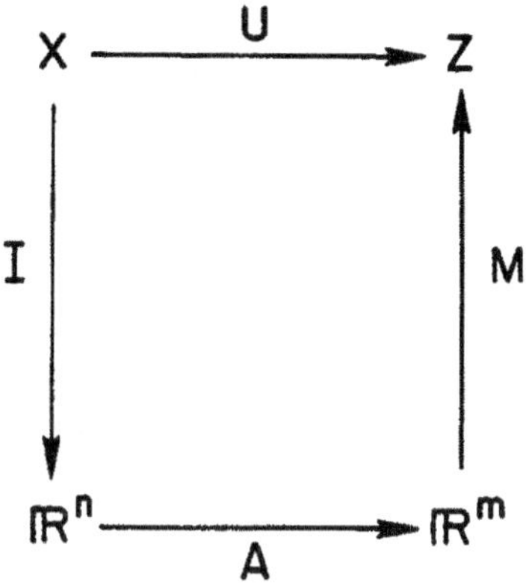

Fig. 5.1

The error produced by this algorithm is

$$\sup_{\|x\| \le 1} \| Ux - MAIx \| ,$$

and an A which minimizes this error determines an optimal algorithm. A linear algorithm corresponds to an $m \times n$ matrix.

Theorem 5.1. Let X, Z be Hilbert spaces and $U : X \to Z$, $I : X \to \mathbb{R}^n$, $M : \mathbb{R}^m \to Z$ be bounded linear operators, and suppose $\varepsilon = 0$. Then there exists a linear optimal algorithm.

Proof. Let A be any mapping from $\mathbb{R}^n$ to $\mathbb{R}^m$. Then the analogue of Theorem 2.1 gives

$$\sup_{\|x\| \le 1} \| Ux-MAIx \| \ge \sup \{ \| Ux \| : \|x\| \le 1, Ix = 0 \} .$$

But we also, obviously, have

$$\sup_{\|x\| \le 1} \| Ux - MAIx \| \ge \sup_{\|x\| \le 1} \inf_{z \in \mathcal{U}} \| Ux - z \|$$

Hence for all A

$$\text{(5.1)} \qquad \sup_{\|x\|\le 1} \|Ux - MAIx\| \ge$$

$$\max\left(\sup\{\|Ux\| : \|x\| \le 1, Ix = o\},\ \sup_{\|x\|\le 1} \inf_{z\in\mathcal{U}} \|Ux - z\|\right).$$

Weinberger [53] showed that there is a matrix which produces an error equal to this lower bound. This result is extended in Davis, Kahan and Weinberger [9]. We follow Micchelli [33] here and present a proof which uses the Kolmogorov characterization of a best approximation in the max norm from a finite dimensional space. Thus we consider the problem

$$\min_{a_{ij}} \sup_{\|x\|\le 1} \|Ux - \sum_{i,j} u_i\, a_{ij}\, (v_j,x)\| .$$

We only discuss the case that U is a compact operator. The general case can be handled similarly. This condition allows the following formulation of the problem. Let $B_X = \{x \in X : \|x\| \le 1\}$ and $B_Z = \{z \in Z : \|z\| \le 1\}$. B_X and B_Z are compact in the weak topology on X and Z respectively. Therefore $T = B_X \times B_Z = \{t = (x,z) : \|x\| \le 1, \|z\| \le 1\}$ is compact in the product topology. For each $t \in T$ we define $f(t) = (Ux,z)$. Then f is continuous on T. We define $(u\otimes v)x = u(v,x)$ so that $u\otimes v : X \to Z$. For every linear operator $S : X \to Z$ we define $g_S(t) = (Sx,z)$, put $\mathcal{N} = \{g_S : S \in \mathcal{U}\otimes\mathcal{V} = \text{span}\{u_i\otimes v_j\}\}$ and write $\mathcal{N} = \text{span}\{w_1(t),\dots,w_N(t)\}$. Then

$$\min_{a_{ij}} \max_{\|x\|\le 1} \|Ux - \sum_{i,j} u_i\, a_{ij}\, (v_j,x)\|$$

$$= \min_{a_{ij}} \max\{\,|(Ux,z) - \sum_{i,j} a_{ij}(v_j,x)(u_i,z)| : \|x\| \le 1,\ \|z\| \le 1\}$$

$$= \min_{c_i} \max_{t\in T} |f(t) - \sum_{i=1}^{N} c_i w_i(t)| ,$$

a standard best approximation problem.

Let the best approximation be $g_0 = g_{S_0} \in \mathcal{N}$. g_0 is characterized (cf. Cheney [7]) by the condition that there are $t_1,\dots,t_p \in T$, $p \le N+1$, with

$$f(t_i) - g_0(t_i) = \max_{t\in T} |f(t) - g_0(t)|$$

and $\mu_i > 0$,

$$\sum_{i=1}^{p} \mu_i = 1$$

such that

(5.2) $$\sum_{j=1}^{p} u_j \, g_{(u_s \otimes v_r)}(t_j) = 0 \qquad s = 1,\ldots,m \; ; \; r = 1,\ldots,n \; .$$

Let $E = U - S_0$, then for $t_i = (x_i, z_i) \in T$ we have $(E\, x_i, z_i) = \|E\|$. It is well-known that this condition is equivalent to the requirement that $E^*Ex_i = \lambda x_i$, $\lambda = \|E\|^2$, $\|x_i\| = \|z_i\| = 1$ and $z_i = Ex_i / \|Ex_i\|$. Hence equations (5.2) simplify to

(5.3) $$\sum_{j=1}^{p} \mu_j (u_s, Ex_j)(v_r, x_j) = 0, \qquad s = 1,\ldots,m \; ; \; r = 1,\ldots,n.$$

Let

$$\zeta_r = \sum_{j=1}^{p} \mu_j (v_r, x_j) x_j$$

then (5.3) becomes

$$(E\zeta_r, u_s) = 0 \; , \qquad s = 1,\ldots,m \; ; \; r = 1,\ldots,n \; .$$

Each ζ_r satisfies $\|E\zeta_r\| = \|E\| \; \|\zeta_r\|$ and thus for any $u \in \mathscr{U}$
$\|E\zeta_r\|^2 = (E\zeta_r, E\zeta_r) = (E\zeta_r, U\zeta_r - u) \le \|E\zeta_r\| \; \|U\zeta_r - u\|$. Furthermore

$$\begin{aligned} \|E\|^2 \; \|\zeta_r\|^2 &\le \|E\zeta_r\| \; \mathrm{dist}\,(U\zeta_r, \mathscr{U}) \\ &= \|E\zeta_r\| \; \max\,\{(U\zeta_r, \omega) : \; \|\omega\| \le 1, \; \omega \in \mathscr{U}^{\perp}\} \\ &= \|E\zeta_r\| \; \min_{z \in \mathscr{U}} \; \|U\zeta_r - z\| \; . \end{aligned}$$

Hence if for some r, $\zeta_r \neq 0$ we obtain

(5.4) $$\|E\| \le \sup_{\|x\| \le 1} \; \inf_{z \in \mathscr{R}(M)} \|Ux - z\|$$

while otherwise we have

$$0 = (v_r, \zeta_r) = \sum_{j=1}^{p} \mu_j (v_r, x_j)^2 \; .$$

Consequently $(v_r, x_j) = 0$, $r = 1,\dots,n$; $j = 1,\dots,p$, and

$$\|E\| = \|Ex_1\| = \|Ux_1\| \le \sup\{\|Ux\| : \|x\| \le 1, Ix = 0\} . \tag{5.5}$$

(5.4) and (5.5) imply that $\|E\|$ does not exceed the lower bound given in (5.1). The theorem is proved. More details about the linear optimal algorithms may be found in Davis, Kahan and Weinberger [9].

Example 5.1. $X = Z = \tilde{L}^2[0,1]$ (one-periodic functions), $Uf = f$, $m < n$,

$$M(b_{-m},\dots,b_m)(t) = \sum_{-m}^{m} b_j e^{2\pi ijt} ,$$

$If = (\hat{f}_{-n},\dots,\hat{f}_n)$ and $K = \{f \in X : \|f^{(k)}\|_2 \le 1\}$. (i.e., in our prior discussion m is replaced by 2m+1, n by 2n+1, $u_j = e^{2\pi ijt}$, $|j| \le m$, $v_j = e^{-2\pi ijt}$, $|j| \le n)$. Then

$$E f(t) = f(t) - \sum_{\ell=-m}^{m} \sum_{r=-n}^{n} e^{2\pi i\ell t} \hat{f}_r a_{\ell r}$$

and

$$\|Ef\|^2 = \sum_{|\ell|\le m} |\hat{f}_\ell - \sum_{|r|\le n} \hat{f}_r a_{\ell r}|^2 + \sum_{|\ell|>m} |\hat{f}_\ell|^2 . \tag{5.6}$$

Note that $f \in K$ implies $\Sigma j^{2k} |\hat{f}_j|^2 \le (4\pi^2)^{-k}$.

Consider the two extremal problems on the right-hand side in (5.1). In the present context they are:

(i) $\sup\{\|f\| : \hat{f}_j = 0, |j| \le n ; \|f^{(k)}\| \le 1\}$ and

(ii) $\sup\{\inf_{\tau\in\mathscr{T}_m} \|f - \tau\|_2 : \|f^{(k)}\| \le 1\}$,

where $\mathscr{T}_m$ is the trigonometric polynomials of degree at most m. In the case of (i) we have

$$(4\pi^2)^{-k} \ge \sum_{|j|>n} j^{2k} |\hat{f}_j|^2 \ge (n+1)^{2k} \sum_{|j|>n} |\hat{f}_j|^2 ,$$

with equality possible. Thus (i) yields $(2\pi(n+1))^{-k}$, while in the case of (ii), since

$$\inf_{\tau\in\mathscr{T}_m} \|f-\tau\|^2 = \sum_{|j|>m} |\hat{f}_j|^2 ,$$

a similar argument gives the answer $(2\pi(m+1))^{-k}$. Since $m < n$ the lower bound on the right-hand side of (5.1) is $(2\pi(m+1))^{-k}$.

Now choose $a_{\ell,r} = 0$, $m < |r| \leq n$ and $a_{\ell,r} = \sigma_\ell \, \delta_{\ell,r}$ where $(1-\sigma_\ell)^2 \leq \ell^{2k}(m+1)^{-2k}$, $|\ell| \leq m$. Then the right-hand side in (5.6) becomes

$$\sum_{|\ell|\leq m} (1-\sigma_\ell)^2 \, |\hat{f}_\ell|^2 + \sum_{|\ell|>m} |\hat{f}_\ell|^2 .$$

Therefore,

$$(4\pi^2)^{-k} \geq \Sigma j^{2k} \, |\hat{f}_j|^2 \geq (m+1)^{2k} \, [\sum_{|\ell|\leq m} (1-\sigma_\ell)^2 \, |\hat{f}_\ell|^2 + \sum_{|\ell|>m} |\hat{f}_\ell|^2] ,$$

and we see that our choice of the $a_{\ell,r}$ provides us with many linear optimal algorithms.

6. Optimal Information

6.1 Sequential Selection of Information

As a means of introducing some results on optimal information let us recall a well-known result about the bisection algorithm for finding a zero of a function. It is known that by evaluating a function n times in [0,1] sequentially, i.e., the choice of each point at which to make the evaluation depends on the previous evaluations, its zero can be located within an interval of length 2^{-n}. However if all n evaluations are performed simultaneously the best that can be done is $1/n$.

For the purpose of determining when sequential use of information is not advantageous we follow Gal and Micchelli [15] in introducing some terminology consistent with that which we have been using. Let K be a set of real-valued functions on [0,1] and U be a functional, $U : K \to \mathbb{R}$. If $f \in K$ is sampled at $x : (x_1,\ldots,x_n)$, $x_i \in [0,1]$ we obtain the information $(f(x_1),\ldots,f(x_n))$. If we put $K_f = \{g \in K : g(x_i) = f(x_i),\ i = 1,\ldots,n\}$ then

$$H(x, f) = \{Uh : h \in K_f\}$$

is the set of uncertainty. As a measure of the size of H we define

$$g(x, f) = \sup_{h \in K_f} Uh - \inf_{h \in K_f} Uh,$$

so that g(x, f) is the length of the smallest interval containing H(x, f). Let $B_n = [0,1]^n$ and consider the following two methods of selecting $x \in B_n$ (a third is considered in Gal and Micchelli [15]).

(i) Sequential

A sequential search procedure is a set of n functions $u_1,...,u_n$, $u_i : \mathbb{R}^{2(i-1)} \to [0,1]$, $i = 1,...,n$, where $x_1 = u_1$ is a constant, $x_2 = u_2(x_1, f(x_1))$,..., $x_{i+1} = u_{i+1}(x_1, f(x_1),...,x_i, f(x_i)),..., x_n = u_n(x_1, f(x_1),...,x_{n-1}, f(x_{n-1}))$. This procedure produces $x = u(f) = (u_1,...,u_n) \in B_n$. The totality of sequential procedures is denoted by S_n. The intrinsic error associated with sequential procedures is given by

$$s_n = \inf_{u \in S_n} \sup_{f \in K} g(u(f), f)$$

(ii) Simultaneous

In this case the intrinsic error is

$$d_n = \inf_{x \in B_n} \sup_{f \in K} g(x, f).$$

Example 6.1. Let $K = \{f : f(0) = -1, f(1) = 1, f(a) \geq 0 \Rightarrow f(b) > 0, a < b\}$ and $Uf = \inf \{t : f(t) > 0\}$. Thus Uf is the "zero" of f. The bisection algorithm shows that $s_n = 2^{-n}$. However, if we put $x_0 = 0$, $x_{n+1} = 1$ then $g(x, f) = x_j - x_{j-1}$ where j is the smallest index such that $f(x_j) > 0$. Hence

$$d_n = \min_{x \in B_n} \max_j (x_j - x_{j-1}) = \frac{1}{n+1}.$$

In contrast to this example the following, among other facts, is established in Gal and Micchelli [15].

Theorem 6.1. Let K be a convex balanced set and U a linear functional, then $d_n = s_n$.

Intuitively, sequential procedures will work better than simultaneous ones if something is "learned" about the unknown function $f \in K$ so that the set of uncertainty is sequentially diminished. However, when the hypothesis of Theorem 6.1 holds it is easy to see that $f = 0$ is a "worst" function for any $x \in B_n$ and so nothing is gained with a sequential strategy.

The remarks above remain valid in more general situations. However, in view of Theorem 6.1, and others given by Gal and Micchelli, we propose next to turn our attention to the question of optimal information in the usual (simultaneous) setting. We shall introduce this problem in a general setting in normed linear spaces. Although the ideas developed are related to the notion of n-widths, this concept will not be introduced. In a later subsection we will summarize some results about optimal procedures for estimating certain function classes.

6.2 Optimal Linear Information

Let X be a normed linear space and K a subset of X. We define

$$i_n(K,X) = \inf_{I,\alpha} \sup_{x \in K} \| x - \alpha\, Ix \|$$

where the infimum is taken over all continuous linear maps $I : X \to \mathbb{R}^n$ and mappings $\alpha : IK \to X$ (not necessarily continuous or linear), i.e., $Y = \mathbb{R}^n$ and $Z = X$ in our usual notation. If I_0 achieves the infimum it is called the optimal linear information operator. We present only one result, which is widely known, about $i_n(K,X)$.

Theorem 6.2. Let W, X be Hilbert spaces and T be a compact linear operator from W into X. Put $K = \{Tw : \|w\| \le 1\}$, then $i_n(K, X) = \lambda_n^{1/2}$ where $\lambda_0 \ge \lambda_1 \ge \lambda_2 \ge \ldots$ are the eigenvalues of $T^*T : W \to W$. Furthermore, if $T^*Tw_i = \lambda_i w_i$, $(w_i, w_j) = \delta_{ij}$, and we put $x_i = Tw_i$, $i = 0,\ldots,n-1$, then $I_0 x = ((x_0,x),\ldots,(x_{n-1},x))$ is optimal linear information.

Proof. In view of Theorem 3.1 we have

$$i_n(K, X) = \inf_I \sup\{ \|x\| : x \in K, Ix = 0\},$$

and so the result follows from the min-max characterization of eigenvalues. Specifically, for every $w \in W$ we have

$$\|Tw\|^2 = \sum_{i=0}^{\infty} \lambda_i \, |(w_i,w)|^2$$

and so

$$\begin{aligned}[i_n(K, X)]^2 &\le \sup\{ \|Tw\|^2 : \|w\| \le 1, (Tw_i, Tw) = 0, i = 0,\ldots,n-1\} \\ &\le \lambda_n .\end{aligned}$$

Moreover, if $u_0,\ldots,u_{n-1}$ are any points of X then there is a vector $(\mu_0,\ldots,\mu_n) \in \mathbb{R}^{n+1}$ such that

$$\tilde{w} = \sum_{i=0}^{n} \mu_i w_i$$

satisfies $\|\tilde{w}\| = 1$ and $(T\tilde{w}, u_i) = 0$, $i = 0,\ldots,n-1$. Hence

$$\sup\{\|Tw\|^2 : \|w\| \le 1, (Tw, u_i) = 0, i = 0,\ldots,n-1)\} \tag{6.1}$$
$$\ge \|T\tilde{w}\|^2 = \sum_{i=0}^{n} \lambda_i \mu_i^2 \ge \lambda_n .$$

The theorem is proved.

Actually, the lower bound result (6.1) can be extended to some cases of nonlinear information by using the Borsuk antipodality theorem (cf. Berger and Berger [3]). Namely, if IT is an odd map which is continuous on the subspace

$$\{\sum_{i=0}^{n} \mu_i w_i\}$$

then there is a $\tilde{w}$ satisfying $\|\tilde{w}\| = 1$ and $IT\,\tilde{w} = 0$. Hence

$$\sup\{\|x\|^2 : x \in K, Ix = 0\} \ge \|T\tilde{w}\|^2 \ge \lambda_n ,$$

and I_0x remains optimal for nonlinear information of the kind we have specified here.

Example 6.2. As an example of Theorem 6.2 let $W = X = \tilde{L}^2\,[0,1]$ (one-periodic complex-valued L^2 functions on [0,1]). For some $\varphi \in X$ let

$$(Tx)(t) = \int_0^1 \varphi(t-s)x(s)ds.$$

Then $|\hat{\varphi}(k)|^2$ are the eigenvalues of T^*T with corresponding orthonormal eigenfunctions $e^{2\pi ikt}$. Since $|\hat{\varphi}(k)| \to 0$ as $k \to \infty$ we can reorder this sequence into a non-increasing sequence and put $\{|\hat{\varphi}(k)|^2 : k = 0, \pm 1,\ldots\} = \{\lambda_0, \lambda_1,\ldots : \lambda_0 \ge \lambda_1 \ge \ldots\}$. Suppose $\lambda_j = |\hat{\varphi}(k_j)|^2$, then optimal information is $\hat{x}_{k_0},\ldots,\hat{x}_{k_{n-1}}$ and the corresponding intrinsic error is $|\hat{\varphi}(k_n)|$.

6.3 Optimal Sampling

In the present setting X is a normed linear space of real-valued (complex-valued) functions on some set B. For a subset K of X and $n \geq 1$ we define

$$i_n(K,X) = \inf_{I,\alpha} \sup_{f \in K} \|f - \alpha If\|, \tag{6.2}$$

where the infimum is taken over all continuous linear maps $I : X \to \mathbb{R}^n$ ($\mathbb{C}^n$) and mappings $\alpha : IK \to X$. We will say that *sampling is optimal for K in X* if there exist n points $t_1,\ldots,t_n \in B$ such that for $I_n f = (f(t_1),\ldots,f(t_n))$

$$i_n(K, X) = \inf_{\alpha} \sup_{f \in K} \|f - \alpha I_n f\| . \tag{6.3}$$

In this case $t_1,\ldots,t_n$ are called *optimal sample points*. When there is a constant, c, such that for all n sufficiently large

$$\inf_{\alpha} \sup_{f \in K} \|f - \alpha I_n f\| \leq c\, i_n(K,X) \tag{6.4}$$

then *sampling is asymptotically optimal for K* in *X*.

All of the material below (as well as (6.2) itself) is related to n-widths of various kinds (Kolmogorov, Gel'fand, Bernstein, Linear, ...). However we will not present these relationships. The relevant information may be found in the original sources for the results quoted below.

(a) Smooth Functions in Sobolev Spaces

Let $W_p^r[0,1]$ be the class of real-valued functions on [0,1] defined by

$$W_p^r[0,1] = \{f : f^{r-1} \text{ absolutely cont., } f^{(r)} \in L^p[0,1]\} , \tag{6.5}$$

and set

$$B_p^r[0,1] = \{f \in W_p^r[0,1] : \|f^{(r)}\|_p \leq 1\}. \tag{6.6}$$

We now describe certain classes of indices p, q, r (q not necessarily the conjugate index to p) $1 \leq p \leq \infty$, $1 \leq q \leq \infty$, $r \geq 1$ for which sampling is optimal for $B_p^r[0,1]$ in $L^q[0,1]$.

For $r = 1$ and $p \geq q$ Makovoz [25] showed that equally spaced points are optimal sample points. This extended earlier work of Babadjanov and Tichomirov [2] for the case $p = q$. It is conjectured (Pinkus [45]) that sampling is optimal for any $r \geq 1$ and $p \geq q$. This has been established in the case $p = q = 2$ in Melkman and Micchelli [28]; $p = \infty$ or $q = 1$ in Micchelli and Pinkus [35]; and recently for $p = q$ in Pinkus [45]. Various extensions of these results for totally positive kernels and classes of periodic functions have been made. We do not consider these extensions here. Instead we now describe the rather involved construction needed to identify optimal sampling points in the cases mentioned above. (In general we need $n \geq r$ sample points because the semi-norm $\| f^{(r)} \|_p$ is zero whenever f is a polynomial of degree $< r$.)

(i) $p = \infty, 1 \leq q \leq \infty$.

We recall some properties of perfect splines. A perfect spline of degree r with k knots $\xi_1,...,\xi_k$ is a function of the form

$$P(x) = \sum_{i=0}^{r-1} a_i x^i + \frac{1}{(r-1)!} \sum_{i=0}^{k} (-1)^i \int_{\xi_i}^{\xi_{i+1}} (x-t)_+^{r-1} \, dt, \tag{6.7}$$

($\xi_0 = 0$, $\xi_{k+1} = 1$). There exists a perfect spline of degree r with n-r knots, P_0, which has smallest L^q norm, and P_0 has exactly n zeros in [0,1]. These zeros are optimal sampling points for B^r_∞ [0,1] in $L^q[0,1]$.

(ii) $1 \leq p \leq \infty, q = 1$.

Let Q_0 be a perfect spline of degree r with n knots, satisfying $Q_0^{(i)}(0) = Q_0^{(i)}(1) = 0, i = 0,1,...,r-1$, of minimum $L^{p'}$ norm, $((1/p') + (1/p) = 1)$ among all such perfect splines. Then Q_0 has exactly n knots and these are optimal sampling points for $B^r_p[0,1]$ in $L^1[0,1]$.

(iii) $1 < p = q < \infty, r \geq 2$.

For each n there is a $\lambda_n > 0$ and a unique $f_n \in W^r_p[0,1]$ with n sign changes such that

$$\frac{1}{(r-1)!} \int_0^1 (x-y)_+^{r-1} \, | f_n(x)) |^{p-1} \operatorname{sgn} f_n(x)dx = \lambda_n | f_n^{(r)}(y) |^{p-1} \operatorname{sgn} f_n^{(r)}(y) \tag{6.8}$$

and

$$\int_0^1 x^i | f_n(x) |^{p-1} \operatorname{sgn} f_n(x)dx = 0, \; i = 0,1,...,r-1. \tag{6.9}$$

The n zeros of f_n are optimal sampling points for $B^r_p[0,1]$ in $L^p[0,1]$.

If we put

$$g_n(y) = \frac{1}{(r-1)!} \int_0^1 (x-y)_+^{r-1} | f_n(x) |^{p-1} \operatorname{sgn} f_n(x)dx \tag{6.10}$$

then $g_n^{(r)}(x) = (-1)^r | f_n(x) |^{p-1} \operatorname{sgn} f_n(x)$ so that (6.9) and (6.8) are equivalent to

$$g_n^{(i)}(0) = g_n^{(i)}(1) = 0, \; i = 0,...,r-1 \tag{6.11}$$

and

$$(6.12)\qquad (-1)^r \lambda_n^{p'-1} \frac{d^r}{dx^r}[\,|g_n^{(r)}(x)|^{p'-1} \operatorname{sgn} g_n^{(r)}(x)] = |g(x)|^{p'-1} \operatorname{sgn} g(x).$$

For p =2 (6.12) and (6.11) become

$$(6.13)\qquad (-1)^r \lambda_n g_n^{(2r)}(x) = g_n(x)\ ;\ g_n^{(i)}(0) = g_n^{(i)}(1) = 0,\ i = 0,1,\ldots,r-1.$$

This eigenvalue problem has a Green's function which is an oscillation kernel (see Gantmacher and Krein [16] and Melkman and Micchelli [28]) and so g_n has n zeros. Successive differentiation substantiates the claim that in this case f_n has n zeros too. The role of these zeros in optimal sampling was observed in Melkman and Micchelli [28]; the general case given by (6.12) and (6.11) is due to Pinkus [45].

In view of the difficulty of identifying optimal sample points it is comforting to know that much less effort is required to show that equally spaced points are *asymptotically optimal* for any $r \geq 1$, $p \geq q$. We now present the proof of this result.

The precise statement is that there exist positive constants c and d such that

$$(6.14)\qquad i_n(B_p^r[0,1], L^q[0,1]) \geq \frac{c}{n^r}$$

and

$$(6.15)\qquad \inf_{\alpha} \sup_{\|f^{(r)}\|_p \leq 1} \|f - \alpha(f(\tfrac{1}{n+1}),\ldots,f(\tfrac{n}{n+1}))\|_q \leq \frac{d}{n^r}.$$

The upper bound, (6.15), is easily proved by using a local polynomial interpolation scheme. Thus we divide [0,1] into m equally spaced intervals. On each of these intervals we interpolate f at r equally spaced points by a polynomial of degree at most r-1. If we now choose m so that n is, roughly, mr (6.15) follows. Actually, one can smooth this approximation by using a spline of degree r-1 with simple knots which are equally spaced; or even use just one polynomial on all of [0,1], by means of Jackson's theorem (cf. Höllig [17′]). Since our algorithm clearly produces a function in $L^q[0,1]$ we do not pursue these alternatives further here.

For the lower bound we follow the method used in Makozov [25]. Let φ be any C^∞ function with support on (0,1) normalized so that $\|\varphi^{(r)}\|_p = 1$. Choose any

$t = (t_1,\ldots,t_{n+1}) \in \mathbb{R}^{n+1}$ such that

$$\sum_{i=1}^{n+1} |t_i| = 1 . \tag{6.16}$$

Define

$$f(x,t) = \sum_{i=1}^{n+1} t_i |t_i|^{r-1} \varphi\left(\frac{x - x_{i-1}}{|t_i|}\right), \tag{6.17}$$

where

$$x_i = \sum_{j=1}^{i} |t_j|,\ i = 1,\ldots,n+1 ;\ x_0 = 0 . \tag{6.18}$$

Let I be any continuous map from $W_p^r[0,1]$ into $\mathbb{R}^n$. Then $t \to I(f(\cdot,t))$ is an odd continuous map and so by the Borsuk antipodality theorem (cf. Berger and Berger [3]) there is a $t^0 = (t_1^0,\ldots,t_{n+1}^0)$, $\Sigma |t_i^0| = 1$, such that $I(f(\cdot, t^0)) = 0$. Now in view of (6.16), (6.17) and (6.18) we have $\|f^{(r)}(\cdot,t)\|_p = 1$ and

$$\|f(\cdot,t^0)\|_q = \left(\sum_{i=1}^{n+1} |t_i^0|^{qr+1}\right)^{1/q} \|\varphi\|_q \geq \frac{1}{(n+1)^r} \|\varphi\|_q,$$

the inequality following from the convexity of x^{1+qr} on [0,1]. Thus we have

$$\begin{aligned} i_n(B_p^r[0,1], L^q[0,1]) &\geq \inf_I \sup \{\|f\|_q : \|f^{(r)}\|_p \leq 1, If = 0\} \\ &\geq \frac{\|\varphi\|_q}{(n+1)^r} , \end{aligned}$$

thereby establishing our claim. We have shown that equally spaced points are asymptotically optimal, but it is not difficult to see that other choices of sampling points would also satisfy (6.15) and hence be asymptotically optimal.

One might expect equally spaced points to be asymptotically optimal in recovering periodic functions. The following function classes were considered in Dahmen, Micchelli and Smith [8]. Let X be $\tilde{L}^2[0,1]$, complex-valued one-periodic functions. For every $\varphi \in X$ we define $K_\varphi = \{\varphi * h : h \in X ,\ \|h\| \leq 1\}$, where * is convolution. There are functions $\varphi \in X$ such that equally spaced sampling is not asymptotically optimal for K_φ in X (cf. Dahmen, Micchelli and Smith [8]). It is not known in those cases whether any sampling is asymptotically optimal. However, whenever there exist constants, L, U, V, and a positive increasing sequence $\Psi(n)$ such that

$$0 < L \leq |\hat{\varphi}(\pm n)|^2 \, \Psi(n) \leq U, \; n = 0,1,2,...,$$

and

$$\overline{\lim_{n \to \infty}} \; \Psi(n) \sum_{j=1}^{\infty} \frac{1}{\Psi(jn)} = V$$

then equally spaced points are asymptotically optimal sampling points for K_φ in X.

(b) Time and Band Limited Functions

Let $L_C^2(\mathbb{R})$ denote complex-valued square integrable functions on $\mathbb{R}$. For any $T > 0$ we suppose that D is the set of all functions in $L_C^2(\mathbb{R})$ vanishing outside $(-T, T)$ (time limited). B is the set of functions, f, ϵ $L_C^2(\mathbb{R})$ such that

$$\hat{f}(t) = \int_{-\infty}^{\infty} e^{2\pi i t \tau} f(\tau) d\tau$$

vanishes outside of $(-\sigma,\sigma)$ (band limited).

Then Melkman [30] proved that sampling is optimal for either

$$\{f \in L^2[-T,T] : f \in B, \int_{-\infty}^{\infty} |f(t)|^2 dt \leq 1\}$$

or

$$\{f \in L^2[-T,T] : f \in B, \int_{|t| \geq T} |f(t)|^2 dt \leq 1\}$$

in $L^2[-T,T]$. Melkman [30] also shows that the optimal points are the same in both cases. To describe this part of his result we introduce orthogonal projections P_D, P_B onto D and B respectively,

$$(P_D f)(t) = \begin{cases} f(t), & |t| \leq T \\ 0, & |t| > T \end{cases},$$

and

$$(P_B f)(t) = \int_{-\infty}^{\infty} f(\tau) \frac{\sin 2\pi\sigma(t-\tau)}{\pi(t-\tau)} dt .$$

Let Ψ_i, λ_i be the orthonormal eigenfunctions and eigenvalues, respectively, of the compact, positive definite and symmetric integral operator

$$(P_B P_D f)(t) = \int_{-T}^{T} f(\tau) \frac{\sin 2\pi\sigma(t-\tau)}{\pi(t-\tau)} d\tau .$$

It is known that Ψ_n has exactly n zeros, $\xi_1,\ldots,\xi_n \in (-T,T)$. This is a consequence of the observation that $P_B P_D$ commutes with the Sturm-Liouville differential operator (prolate spheroidal wave equation)

$$(Lf)(t) = ((T^2-t^2)f'(t))' - 4\pi^2\sigma^2 t^2 f(t).$$

Thus the Ψ_i, being the (regular) eigenfunctions of Lf, have i distinct zeros and the λ_i are distinct. With these facts at hand Melkman proved that $\xi_1,\ldots,\xi_n$ are optimal sampling points.

He has also provided an analogue of this result in the sup-norm. Namely let $K(\sigma, T)$ be the set of entire functions of exponential type σ which are bounded by 1 on $|t| > T$. Melkman [30] now proves that sampling is optimal for $K(\sigma, T)$ in $C[-T,T]$. As previously the optimal sampling points can be described as zeros of an extremal function in $K(\sigma, T)$ which is defined as follows. Given any n there is a unique real-valued $f_n \in K(\sigma,T)$ which equioscillates exactly n + 1 times at points $x_1,\ldots,x_{n+1}$ of $[-T,T]$, i.e., $f_n(x_i) = (-1)^{n+1-i} \|f_n\|_T$. Moreover, f_n equioscillates outside $(-T,T)$. Also if $\|f_n\|_T < 1$ then $|f_n(\pm T)| = \|f_n\|_T$ and otherwise $|f_n(\pm T)| = 1$. This function has n simple zeros in $(-T,T)$ and it is these zeros that are optimal sampling points for $K(\sigma,T)$ in $C[-T,T]$.

(c) Analytic Functions on the Disc

The setting for this investigation is a domain, Ω, in the complex plane and a compact subset, G, of Ω. For $1 \le p \le \infty$, $H^p(\Omega)$ is the Hardy space on Ω and $A_p(\Omega)$ denotes the unit ball in $H^p(\Omega)$. Let $d\mu$ be a positive measure on G. Suppose X to be $H^p(\Omega)$ normed by its restriction to $L^q(G,d\mu)$ or $C(G)$, i.e., for $f \in H_p(\Omega)$ either

$$\|f\| = [\int_G |f|^q d\mu]^{1/q}, \ 1 \le q < \infty$$

or

$$\|f\| = \max_{z\in G} |f(z)|.$$

Let $K \subset X$ be $A_p(\Omega)$.

Fisher and Micchelli [12] showed that when $\Omega = D = \{z : |z| < 1\}$ sampling is optimal for $A_\infty(D)$ in X. Furthermore, if we recall that a *Blaschke product* of degree m is a function of the form

$$B(z) = \lambda \prod_{j=1}^{m} \frac{z-\alpha_j}{1-\bar{\alpha}_j z}, \quad |\alpha_j| < 1, j = 1,...,m \; ; \; |\lambda| = 1$$

(which is clearly in $A_\infty(D)$), the n optimal sampling points are the zeros of the Blaschke product of degree n which has least norm as an element of X. It could happen that the minimal Blaschke product has multiple zeros in which case we interpret "function values" to mean values of consecutive derivatives equal in number to the multiplicity of the zero.

In Fisher and Micchelli [13] it was shown that sampling is optimal for $A_2(D)$ in X (normed by its restriction to $L^2(G, d\mu)$) . Here the optimal sampling points are the zeros of the Blaschke product of degree n which minimizes

$$\max_{f \in H^2(D)} \frac{\int_G |f(z)|^2 |B(z)|^2 d\mu(z)}{\frac{1}{2\pi}\int_{-\pi}^{\pi} |f(e^{i\theta})|^2 d\theta} .$$

As described earlier in the real case, $p = q = 2$, and the band/time-limited function case, the zeros of the minimizing Blaschke product are related to the zeros of eigenfunctions of a certain integral operator. Recently this result has been extended from $H^2(D)$ to a class of reproducing kernel Hilbert spaces of the following form. Let $0 < \beta_0 \le \beta_1 \le \dots$ be an increasing sequence satisfying

$$\lim_{n \to \infty} \beta_n^{1/n} = 1 .$$

We define H to be the Hilbert space of all functions $f = \Sigma a_n z^n$ holomorphic in D and such that

$$\|f\|_H^2 = \sum_{n=0}^{\infty} |a_n|^2 \beta_n \le \infty .$$

Then sampling is again optimal for the unit ball of H restricted to G in X (normed by its restriction to $L^2(G, d\mu)$), and the optimal sampling points are the zeros of the Blaschke product of degree n which minimizes

$$\max_{f\in H} \frac{\int_G |f(z)|^2 |B(z)|^2 d\mu(z)}{\|Bf\|_H^2} .$$

7. Stochastic Optimal Recovery

So far we have always assumed that any element $x \in K$ had an equally likely chance to be chosen when estimating Ux. This led us to the worst case error criterion (for exact information), given in Section 2, which required us to take the supremum of $\|Ux-\alpha\, Ix\|$ over $x \in K$. In this section we present a model which allows some randomness in the choice of x. This is accomplished by using a probability measure on X, as in Larkin [20]. Here, however, we restrict ourselves to exact information. The notation and background material that we require on measures for infinite dimensional spaces is taken from Kuo [19] and the new results we shall present come from the recent paper of Micchelli [34].

As a simple example of what we have in mind in this section we consider the elementary

Example 7.1. (Wiener measure) Let $K = X = \{x \in C\,[0,1] : x(0) = 0\}$, $Ix = (x(t_1),\dots,x(t_n))$, $0 < t_1 <\dots< t_n < 1$ and $Ux = x(t)$ for some $t \in (0,1]$. We wish to estimate Ux optimally by a linear algorithm,

$$\alpha(Ix) = \sum_{i=1}^{n} \alpha_i x(t_i) ,$$

which gives the least mean square error with respect to Wiener measure on X. That is, we seek to minimize

$$\int_X |x(t) - \sum_{i=1}^{n} \alpha_i x(t_i)|^2 W(dx) .$$

If we put

$$K(t,s) = \int_X x(t)x(s)W(dx)$$

then the normal equations for the minimizing $\alpha_1^0(t),\ldots,\alpha_n^0(t)$ are

$$K(t,t_i) = \sum_{j=1}^{n} \alpha_j^{(0)} K(t_i,t_j) \ , \ i = 1,\ldots,n \ .$$

Since $K(t,s) = \min(t,s)$ (See Kuo [19 , p. 38]) and $\alpha_j^0(t_i) = \delta_{ij}$ it follows that for $t \in [t_1,t_n]$ an optimal algorithm

$$(\alpha^0 Ix)(t) = \sum_{i=1}^{n} \alpha_i^0(t)x(t_i)$$

is the piecewise linear interpolant to $x(t_1),\ldots,x(t_n)$ with break-points at $t_1,\ldots,t_n$. For $t > t_n$ it is $x(t_n)$, and when $t < t_1$ it is $(x(t_1)/t_1)t$.

This example extends to the following setting. Let $X = K$ be a Hilbert space and μ a Borel measure defined on X. Suppose $I : X \to \mathbb{R}^n$ is a continuous linear operator and U is a continuous linear operator from X into Z, which is a Hilbert space. We wish to estimate Ux , $x \in X$, by a linear algorithm $\alpha(Ix)$ and find such an algorithm with least mean square error

$$E_\alpha = \int_X \| Ux - \alpha\ Ix \|^2 \mu\ (dx).$$

Note that we may write $Ix = ((x_1,x),\ldots(x_n x))$ where $x_1,\ldots,x_n \in X$. Just as in the example it is easy to see that the normal equations for an optimal algorithm,

$$\alpha^0(\eta_1,\ldots,\eta_n) = \sum_{i=1}^{n} z_i^0\ \eta_i \ , \quad z_1^0,\ldots,z_n^0 \in Z$$

are

(7.1) $$\int_X (x_j,x)\ (Ux,\ z)\mu(dx) = \sum_{i=1}^{n} (z_i^0,z) \int_X (x_i,x)\ (x_j,x)\ \mu(dx), \quad j = 1,\ldots,n,$$

which holds for all $z \in Z$. Assuming, as we do, that

$$\int_X \| x \|^2\ \mu(dx) < \infty$$

we can use the fact that X is a Hilbert space and introduce the covariance operator

$$(S_\mu\ x,y) = \int_X (x,v)(y,v)\mu(dv).$$

Thus S_μ is a positive definite self-adjoint trace class operator, which, for simplicity, we assume to be injective. This means that the range of $\sqrt{S_\mu}$ induces a Hilbert subspace of X, $X_0 = \sqrt{S_\mu}(X)$, with inner product

$$(7.2) \qquad (\sqrt{S_\mu}x, \sqrt{S_\mu}y)_0 = (x,y) \; ; \; x,y \in X.$$

For example, if $X = \mathbb{R}^k$ with the usual inner product

$$(x,y) = \sum_{i=1}^{k} x_i y_i \, ,$$

$x = (x_1,...,x_k)$, $y = (y_1,...,y_k)$, then S_μ is the $k \times k$ matrix whose entries are

$$M_{ij} = \int_X x_i x_j \, \mu \, (dx_1,...,dx_k)$$

and X_0 can be identified with $\mathbb{R}^k$ equipped with a new norm

$$(x,y)_0 = \Sigma M_{ij}^{-1} \, x_i y_j \, .$$

Returning to the general case, we recall that $Ix = ((x_1,x),...,(x_n,x))$, $x_1,...,x_n \in X$ and so its representation in X_0 is $((S_\mu x_1,x)_0,...,(S_\mu x_n,x)_0)$. Now (7.1) yields

$$z_i^0 = \sum_{j=1}^{n} a_{ij} \, U \, S_\mu x_j$$

where $A = (a_{ij})$ is the inverse of the $n \times n$ matrix whose entries are $(S_\mu \, x_i, \, x_j)$, and

$$(7.3) \qquad \alpha^0(Ix) = \sum_{i,j=1}^{n} a_{ij} \, U \, S_\mu \, x_j(x_i,x) \, .$$

(The matrix is invertible because S_μ is injective and $x_1,...,x_n$ are required to be linearly independent.)

If we restrict x to X_0 in (7.3) we can relate our result to that given in Section 3. Namely, we get

$$\alpha^0 I = U \, (\sum_{i,j} g_{ij}^{-1} \, S_\mu \, x_i \otimes S_\mu \, x_j) \, ,$$

$g_{ij} = (S_\mu x_i, \, S_\mu \, x_j)_0 = (S_\mu \, x_i,x_j)$, i.e., the orthogonal projection of X_0 on the range of $(I \,|\, _{X_0})^*$. This is the optimal recovery encountered in Section 3. But note that it is constructed relative to the norm on X_0, rather than relative to the original Hilbert space norm.

These are, trivially, the same when $\dim X < \infty$, and the covariance operator then corresponds to an identity matrix.

We have shown that α^0 is an optimal algorithm in the class of *linear* algorithms for *any* measure. Next we shall show that it is optimal among *all* algorithms, linear or nonlinear, for the class of right unitarily invariant measures introduced by Micchelli [34]. This result is a special case of a general principle, which we will now present, which exhibits parallels between the stochastic and deterministic (worst case) points of view.

In each of these cases we have a functional, H, whose domain, D, are mappings from X into Z. For example,

$$H(U) = \begin{cases} \left[\int_X \|Ux\|^2 \mu(dx)\right]^{1/2}, & \text{(stochastic)}, \\ \sup_{x \in K} \|Ux\|, & \text{(deterministic)}. \end{cases}$$

For general H we wish to minimize $H(U-\alpha I)$ over all mappings $\alpha : Y \to Z$, where $U : X \to Z$, $I : X \to Y$ and $U-\alpha I \in D$. We suppose that $H : D \to \mathbb{R}_+$, and that D is a subspace of the space of all mappings from X into Z. In addition we make the following assumptions about H:

(i) $H(-U) = H(U)$, $U \in D$.

(ii) $H(\frac{1}{2}(U+V)) \leq \max(H(U), H(V))$, (quasi – convexity).

Furthermore, we require the following: There is a quadratic norm on X such that for all isometries $R : X \to X$ (relative to this norm)

(iii) If $U \in D$ then $UR \in D$ and $H(UR) = H(U)$.

We refer to condition (iii) when the field of scalars is complex as *right unitary invariance.*

Keeping in mind that X is equipped with the quadratic norm mentioned in (ii) we define

$$\alpha^0 Ix = UQx$$

where

$$Q = \sum_{i,j} g_{ij}^{-1} x_i \otimes x_j, \quad g_{ij} = (x_i, x_j), \quad i,j = 1,\ldots,n,$$

and observe that $Ix = ((x_1,x),\ldots,(x_n,x))$. Following Micchelli [34] we see that if we put $Px = x - Qx$ then $R = Q - P$ is an isometry satisfying $IR = I$ and $PR = -P$ (generalized Householder transformation). Therefore, if α is any algorithm satisfying $UP - \alpha I \in D$ we have, by (i)

$$H(UP-\alpha I) = H(UPR-\alpha IR) = H(UP + \alpha I) .$$

Hence by (ii)

$$H(UP) \le H(UP-\alpha I) , \tag{7.4}$$

which, upon taking into account a slight abuse of notation involving α, implies that α^0 is optimal among all algorithms.

In view of this general result one might wonder whether the piecewise linear interpolant described in Example 7.1 is optimal among all algorithms using the information. It is not difficult to see that this is the case, as we show next. We assume, for simplicity, that $t_1 < t < t_2$. Other cases can be treated similarly.

Recall that for any measurable function $A(\sigma_1,...,\sigma_n)$ and $0 < \tau_1 <...< \tau_n \le 1$

$$\int_X A(x(\tau_1),...,x(\tau_n))\, W(dx)$$

$$= \frac{1}{((2\pi)^n\, \Delta\tau_1 \dots \Delta\,\tau_n)^{1/2}} \int_{\mathbb{R}^n} A(\sigma_1,...,\sigma_n) e^{-(1/2) \sum_{i=1}^{n} \frac{(\Delta\sigma_i)^2}{\Delta\tau_i}} d\sigma_1 \dots d\sigma_n ,$$

where $\Delta\tau_i = \tau_i - \tau_{i-1}$, $\Delta\sigma_i = \sigma_i - \sigma_{i-1}$, $i = 1,...,n$ $(\sigma_0 = \tau_0 = 0)$. Therefore, if $t_1 < t < t_2$ then

$$\int_X x(t) A(x(t_1), x(t_2))\, W(dx) = \frac{1}{((2\pi)^3 t_1 (t-t_1)(t_2-t))^{1/2}} \times$$

$$\int_{\mathbb{R}^2} A(u_1, u_2) \Big(\int_{\mathbb{R}} u e^{-(1/2)[\frac{u_1^2}{t_1} + \frac{(u-u_1)^2}{t-t_1} + \frac{(u_2-u)^2}{t_2-t}]} du \Big) du_1 du_2$$

$$= \frac{1}{((2\pi)^3 t_1(t-t_1)(t_2-t))^{1/2}} \quad \times$$

$$\int_{\mathbb{R}^2} A(u_1,u_2) \left(\frac{2\pi(t-t_1)(t_2-t)}{t_2-t_1}\right)^{1/2} \left(u_1 \frac{t_2-t}{t_2-t_2} + u_2 \frac{t-t_1}{t_2-t_1}\right) \times$$

$$e^{-(1/2)\left(\frac{u_1^2}{t_1}+\frac{(u_2-u_1)^2}{t_2-t_1}\right)} du_1\, du_2$$

$$= \int_X (\alpha_1^0(t)x(t_1) + \alpha_2^0(t)x(t_2))A(x(t_1), x(t_2))W(dx) .$$

Consequently, for any measurable function, A,

$$\int_X (x(t)-\sum_{i=1}^{n} \alpha_i^0(t)x(t_i))A(x(t_1),\ldots,x(t_n))W(dx) = 0,$$

which is the relation needed to prove the optimality (and uniqueness a.e. W(dx)) of

$$\sum_{i=1}^{n} \alpha_i^0(t)x(t_i)$$

among all algorithms.

It is an easy matter to conclude from this result that for any continuous linear operator U on X

$$\sum_{i=1}^{n} (U\alpha_i^0(t))x(t_i)$$

effects the optimal recovery of Ux, using the information $x(t_1),\ldots,x(t_n)$, among all algorithms using this information.

Returning now to the general case we observe that there are several ways of constructing functionals satisfying conditions (i), (ii) and (iii) above. For instance

$$H(U) = \sup\, \{f(\|Ux\|\ ,\ \|x\|) : x \in X\}$$

has this property if X is a Hilbert space and f(s,t) is convex and nondecreasing in $s \geq 0$ for all $t \geq 0$. Here H is right unitarily invariant relative to the Hilbert space norm on X. The choice $f(s,t) = s/t$ gives the worst case point of view.

Another class of examples is obtained in the following way. Let X be a separable Hilbert space and μ a Gaussian measure on X with mean zero (cf. Kuo [19]), then for any p, $1 \leq p \leq \infty$

$$(7.5) \qquad H_p(U) = \begin{cases} \left(\int_X \|Ux\|^p \mu(dx)\right)^{1/p}, & p < \infty \\ \sup\{\|Ux\| : \|x\| \leq 1\}, & p = \infty . \end{cases}$$

is right unitarily invariant relative to the norm on X_0 defined in (7.2). Therefore, for all p, $\alpha^0 I = UQ$ is an optimal algorithm whether Z is a Hilbert space, or not.

Let us observe that when Z is a Hilbert space $H_p(U)$ has an additional invariance property which will allow us to identify optimal information. It is helpful first to recall the result for $p = \infty$ (deterministic case). Thus X, Z are assumed to be Hilbert spaces and $U : X \to Z$ a compact operator. Optimal information is obtained by minimizing

$$\inf_{\alpha} \sup_{\|x\| \leq 1} \|Ux - \alpha Ix\| = \sup\{\|Ux\| : \|x\| \leq 1, Ix = 0\}$$

over all $I : X \to \mathbb{R}^n$. Just as in Theorem 6.2 it can be shown that optimal information is obtained from

$$(7.6) \qquad I_0 x = ((x_1,x),\ldots,(x_n,x)), \ (x_i,x_j) = \delta_{ij}, \ i, j = 1,\ldots,n,$$

where $U^*Ux_i = \lambda_i x_i$, $\lambda_1 \geq \lambda_2 \geq \ldots \geq 0$. This result holds for all the norms given by (7.5). In fact the optimality of the information (7.6) persists, at least, for all unitarily invariant norms in the sense of Mirsky [40].

Next we denote by $\mathscr{L}(X,Z)$ the set of all bounded linear operators from Hilbert space X to Hilbert space Z. We say that a norm, H, is unitarily invariant with respect to an ordered pair of Hilbert spaces (X,Z) if $H(TU) = H(UR)$ for any isometries T, R and *any mapping* (not necessarily linear in this definition), U, from X to Z. In particular, when μ is a Gaussian distribution on X the family of norms (7.5) are unitarily invariant with respect to (X_0,Z) (Kuo [19]). The following result identifies optimal information for unitarily invariant norms.

Theorem 7.1 (Micchelli [34]). Let H be unitarily invariant with respect to (X,Z). Suppose U is a compact operator with spectral (singular value) decomposition

$$U^*U = \sum_{i=1}^{\infty} \lambda_i x_i \otimes x_i, \ (x_i,x_j) = \delta_{ij}, \ i,j = 1,\ldots,n,$$

$\lambda_1 \geq \lambda_2 > \dots \geq 0$. Then

$$\min_{\alpha,I} H(U-\alpha I) = H(U-\alpha^0 I_0),$$

where

$$\alpha^0 I_0 = UQ_0 \; ; \; Q_0 = \sum_{i=1}^{n} x_i \otimes x_i$$

and

$$I_0 x = ((x_1,x)_0,\dots,(x_n,x)_0) \; .$$

Proof. According to (7.4), if $Q : X \overset{\text{orth}}{\rightarrow} \mathscr{R}(I^*)$ then

$$\min_{\alpha,I} H(U-\alpha I) \geq \min_{I} H(U-UQ)$$

$$\geq \min\{H(U-V) : V \in \mathscr{L}(X,Z), \dim \mathscr{R}(V) \leq n\}.$$

According to Mirsky [40] this lower bound is achieved for the operator

$$V_0 = \sum_{i=1}^{n} U\, x_i \otimes x_i = UQ_0 = \alpha_0 I_0 \, ,$$

which proves the result.

8. Optimal Interpolation of Analytic Functions

Let D be the open unit disc in the complex plane. Suppose $X = H^\infty$, the set of bounded analytic functions in D. If $f \in X$, $\|f\| = \sup\{|f(z)| : z \in D\}$. $K = \{f \in H^\infty : \|f\| \leq 1\}$; $\zeta, z_1, z_2,\dots,z_n \in D$ are given. $Z = \mathbb{C}$, $Uf = f(\zeta)$, $\varepsilon = 0$. If $= (f(z_1),\dots,f(z_n))$, $Y = \mathbb{C}^n$, equipped with the maximum norm. The problem just specified is that of optimal interpolation in H^∞. Note that we do not require that the sampling points, z_i, should be distinct. If some of the sampling points coincide we adopt the convention that the corresponding function values in the definition of If are replaced by the obvious number of consecutive derivatives at the coincident point. Let us now solve this optimal recovery problem.

Consider the Blaschke product

$$B_n(z) = \prod_{i=1}^{n} \frac{z-z_i}{1-\bar{z}_i z} .$$

It is clear that $B_n \in K$ and $B_n(z_i) = 0$, $i = 1,\dots,n$. Therefore (2.1) implies that $E(K,0) \geq |B_n(\zeta)|$. (Indeed, $E(K,0) = |B_n(\zeta)|$, as is easily seen by applying the maximum principle to f/B_n.) Next let the quantities $a_j(\zeta)$, $j = 1,\dots,n$ be determined by the calculus of residues from

$$f(\zeta)-\sum_{j=1}^{n} a_j(\zeta)f(z_j) = \frac{1}{2\pi i}\int_{|z|=1} \frac{B_n(\zeta)}{B_n(z)}\,\frac{1-|\zeta|^2}{1-\bar{\zeta}z}\,\frac{1}{z-\zeta}\,f(z)dz. \tag{8.1}$$

For example, if the z_j are distinct we obtain

$$a_j(\zeta) = \frac{B_n(\zeta)}{B_n'(z_j)}\,\frac{1-|\zeta|^2}{(1-\bar{\zeta}z_j)(\zeta-z_j)} , \quad j = 1,\dots,n,$$

while if $z_1 = z_2 = \dots = z_n$

$$a_{j+1}(\zeta) = (1-|\zeta|^{2(n-j)})\,\frac{\zeta^j}{j!} , \quad j = 0,\dots,n-1 . \tag{8.2}$$

Now suppose $f \in K$. Then (8.1) yields

$$\begin{aligned} |f(\zeta)-\sum_{j=1}^{n} a_j(\zeta)f(z_j)| &\leq |B_n(\zeta)|\,\frac{1}{2\pi}\int_0^{2\pi} \frac{1-|\zeta|^2}{|1-\bar{\zeta}e^{i\theta}|}\,\frac{1}{|e^{i\theta}-\zeta|}\,d\theta \\ &\leq |B_n(\zeta)|\,\frac{1}{2\pi}\int_0^{2\pi} \frac{1-|\zeta|^2}{|e^{i\theta}-\zeta|^2}\,d\theta \\ &\leq |B_n(\zeta)| , \end{aligned}$$

if we recall that $(1-|\zeta|^2)/|e^{i\theta}-\zeta|^2$ is the Poisson kernel. Thus

$$|B_n(\zeta)| \leq E(K,0) \leq \sup_{f\in K} |f(\zeta)-\sum_{j=1}^{n} a_j(\zeta)f(z_j)| \leq |B_n(\zeta)| .$$

Hence

$$\alpha : (f(z_1),\dots,f(z_n)) \to \sum_{j=1}^{n} a_j(\zeta)f(z_j)$$

is a (linear) optimal algorithm, and the intrinsic error in this problem is $|B_n(\zeta)|$. Note that when $z_1 = z_2 = \ldots = z_n = 0$ our optimal algorithm is not the partial sum of the Taylor series evaluated at ζ. There also exist nonlinear optimal algorithms but α can be shown to be the unique linear optimal algorithm (see M-R).

The problem of optimal sampling points in this setting can have meaning if the sampling points, and ζ, are restricted in D. Osipenko [43] considers the case that $\zeta, z_1,\ldots,z_n$ are real and contained in the interval $[a,b] \subset D$ and the optimal recovery of the *function* $f(\zeta)$, $a \leq \zeta \leq b$ is required. He observes that

$$\min_{z_1,\ldots,z_n} \max_{a\leq\zeta\leq b} |B_n(\zeta\,;\ z_1,\ldots,z_n)| = E$$

is attained for

$$z_j = \frac{\mathrm{dn}(\frac{2j-1}{2n}K,k) - \tau(b)}{\mathrm{dn}(\frac{2j-1}{2n}K,k) + \tau(b)}, \quad j = 1,\ldots,n,$$

where dn is a Jacobian elliptic function, $\tau(t) = (1-t)/(1+t)$,

$$K = \int_0^1 \frac{dt}{[(1-t^2)(1-k^2t^2)]^{1/2}},$$

and

$$k = [1 - \frac{\tau^2(b)}{\tau^2(a)}]^{1/2}.$$

The intrinsic error for this choice of sampling points is

$$E = 2^{k+1}h^{-(k+1)n} + O(h^{-(k+5)n}), \quad n \to \infty,$$

where

$$h = e^{\frac{\pi K'}{2K}},$$

and K' is the complete elliptic integral of the first kind with respect to the complementary modulus, k'.

A special case of optimal interpolation in H^∞ with the same setting as given above except that $\varepsilon \geq 0$, that is, with inaccurate data permitted, was studied recently by Osipenko [44]. We next give a brief sketch of these results.

Suppose $-1 < z_1 < z_2 < \dots < z_n < 1$ and $-1 < \zeta < 1$. Recall that in view of Theorem 2.5 we have

$$E(K,\varepsilon) = \sup \{ |f(\zeta)| : f \in K, |f(z_i)| \le \varepsilon, i = 1,\dots,n\} .$$

If $\varepsilon \ge 1$ the optimal recovery problem is trivial. $E(K, \varepsilon) = 1$ and $\alpha : y \to 0$ is optimal. Suppose henceforth that $0 \le \varepsilon < 1$. If $\zeta = z_k$ then $E(K, \varepsilon) = \varepsilon$ and $\alpha : y = (y_1,\dots,y_n) \to y_k$ is optimal. Henceforth we consider only the case $\zeta \ne z_k$, $k = 1,\dots,n$. Osipenko now establishes the following facts.

(i) Suppose $\varepsilon > 0$, $A \subset (-1,1)$ is closed and $\zeta \in (-1,1)\backslash A$. If

$$|f^*(\zeta)| = \sup \{ |f(\zeta)| : f \in K ; |f(z)| \le \varepsilon < 1, z \in A\}$$

then

$$f^*(z) = \lambda \prod_{j=1}^{m} \frac{z-\alpha_j}{1-\bar{\alpha}_j z}, \quad \alpha_j \in (-1,1), \ |\lambda| = 1 .$$

and f^* is the unique extremal, up to the factor λ.

(ii) Normalize f^* by choice of λ so that $f^*(\zeta) > 0$. Then $f^*(\zeta) = E(K, \varepsilon) \Leftrightarrow |f^*(z_i)| \le \varepsilon$ and there exist $z_{i_1} < z_{i_2} < \dots < z_{i_m}$ such that

(8.3) $$f^*(z_{i_k}) = \begin{cases} (-1)^{p+k} \varepsilon, & k = 1,\dots,p \\ (-1)^{p+k+1} \varepsilon, & k = p+1,\dots,m \end{cases}$$

where $\zeta \in (z_{i_p}, z_{i_{p+1}})$ and $\lambda = (-1)^{m+p}$, $(z_{i_0} = -1, z_{i_{m+1}} = 1)$.

As a simple example of (i) and (ii) consider the case $n = 1$, $z_1 = 0$, previously studied in Rivlin [46]. In this case

$$f^*(z) = \operatorname{sgn} \zeta \ \frac{z + \varepsilon \operatorname{sgn} \zeta}{1 + \varepsilon(\operatorname{sgn} \zeta) z},$$

$$E(K, \varepsilon) = \frac{|\zeta| + \varepsilon}{1 + \varepsilon |\zeta|},$$

$m = 1$, $\lambda = \operatorname{sgn} \zeta$ and $\varepsilon = f^*(0)$. Also

$$\alpha : y \to \frac{1-\zeta^2}{1 + \varepsilon\zeta^2} \ y$$

is the unique linear optimal algorithm.

(iii) If $A = \{z_1,\ldots,z_n\}$ in (i) the extremal Blaschke product is of degree $m \leq n$. The sampling points other than $z_{i_1},\ldots,z_{i_m}$ may be disregarded in the optimal recovery problem $\Longleftrightarrow$ (8.3) holds.

An explicit optimal algorithm

$$\alpha : (y_1,\ldots,y_n) \to \sum_{j=1}^{m} \alpha_j \, y_{i_j}$$

is described in Osipenko [44]. If $|f^*(z)| < \varepsilon$ for $z \in \{z_1,\ldots,z_n\} \setminus \{z_{i_1},\ldots,z_{i_m}\}$ then this α is the unique linear optimal algorithm. Osipenko also gives an inductive algorithm for determining a "complete information set" $z_{i_1},\ldots,z_{i_m}$.

(iv) The relationship between the intrinsic error in the case $\varepsilon = 0$ and that in the case $\varepsilon > 0$, for small ε, satisfies

$$E(K, \varepsilon) = E(K,0) + O(\varepsilon^2), \quad \varepsilon \to 0 .$$

This concludes our sketch of Osipenko's treatment of optimal interpolation of H^∞ functions from inaccurate data.

Finally, we return to the original case, $\varepsilon = 0$, but generalize to the setting $X = H^p$, $1 < p \leq \infty$, $K = \{f \in H^p : \|f\|_p \leq 1\}$. Following Fisher and Micchelli [12] we determine $b_1(\zeta),\ldots,b_n(\zeta)$ by the calculus of residues from

$$f(\zeta) - \sum_{j=1}^{n} b_j(\zeta) f(z_j) = \frac{1}{2\pi i} \int_{|z|=1} K_p(\zeta,z) f(z)\,dz$$

where

$$K_p(\zeta,z) = \frac{B_n(\zeta)}{B_n(z)} \; \frac{(1-|\zeta|^2)^{1-\frac{2}{p}}}{(1-\bar{\zeta}z)^{1-\frac{2}{p}}} \; \frac{1}{z-\zeta} .$$

For $f \in K$ Hölder's inequality now yields

$$\left| \frac{1}{2\pi i} \int K_p(\zeta,z) f(z)\,dz \right| \leq \frac{|B_n(\zeta)|}{(1-|\zeta|^2)^{1/p}} .$$

But if

$$g(z) = (1-|\zeta|^2)^{1/p} \, \frac{B_n(z)}{(1-\bar{\zeta}z)^{2/p}}$$

then $g \in H^p$, $\|g\|_p = 1$ and $g(z_i) = 0$, $i = 1,...,n$. Thus

$$\frac{|B_n(\zeta)|}{(1-|\zeta|^2)^{1/p}} = |g(\zeta)| \leq e(K,0) \leq E(K,0) \leq \frac{|B_n(\zeta)|}{(1-|\zeta|^2)^{1/p}} ,$$

and so

$$\alpha : (f(z_1),...,f(z_n)) \to \sum_{j=1}^{n} b_j(\zeta)f(z_j)$$

is a (unique) linear optimal algorithm and its intrinsic error is

$$\frac{|B_n(\zeta)|}{(1-|\zeta|^2)^{1/p}} .$$

Note that when $p = 2$ and $z_1 = z_1 = ... = z_n$ this optimal algorithm produces the $(n-1)^{st}$ partial sum of the Taylor series of f, evaluated at ζ .

9. Optimal Numerical Integration

We presented a simple example of optimal numerical integration in Section 1. This was but a hint of the vast literature on this topic. A fairly recent survey is to be found in Levin and Girshovich [22]. In this section we exhibit the relationship between optimal quadrature (= optimal numerical integration) and optimal monosplines, and mention a few results concerning optimal numerical integration in Sobolev spaces. Then we return to the complex domain and treat numerical integration of analytic functions.

9.1 Monosplines and Numerical Integration in Sobolev Spaces

Recall the notation involving Sobolev spaces defined in (6.5), (6.6). Suppose $1 \leq p \leq \infty$, $r \geq 1$. Consider the following optimal recovery problem. $X = W_p^r[0,1]$, $K = B_p^r[0,1]$. $x = (x_1,...,x_n)$ $(n \geq r)$ is given where $0 \leq x_1 < x_2 <...< x_n \leq 1$. $If = (f(x_1),...,f(x_n))$, $Y = \mathbb{R}^n$, $Z = \mathbb{R}$

$$Uf = \int_0^1 f(t)dt,$$

and $\varepsilon = 0$. The intrinsic error in this problem ($\varepsilon = 0$ being tacitly understood) is denoted by E(K,x). Since our general theory (Section 2) guarantees the existence of a linear optimal algorithm in the present setting we restrict the competition among algorithms to linear ones

$$\alpha : (f(x_1),...,f(x_n)) \to \sum_{i=1}^{n} a_i f(x_i) .$$

Let $a = (a_1,...,a_n)$. In the literature one generally encounters the notation $Qf = \alpha(If)$ and $Rf = Uf - Qf$, so that

(9.1) $$E(K,x) = \inf_{a} \sup_{f \in K} |Rf| .$$

The extremal problem presented in (9.1) is that of determining the optimal weights in a quadrature formula with preassigned nodes. But it is frequently of interest to determine the *best* quadrature formula, namely to minimize $E(K,x)$ over all choices of the nodes x. That is, we seek to determine

$$i_n(K) = \inf_{a,x} \sup_{f \in K} |Rf| ,$$

and the a,x for which this infimum is attained, which then provide a best quadrature formula. The notion of a monospline is important in this effort as we shall see next.

Suppose $\xi = (\xi_1,...,\xi_k)$ where $0 < \xi_1 <...< \xi_k < 1$. Recall that

$$S_{\ell,k}(\xi) = \{s(t) \in C^{\ell-2} (-\infty, \infty) : s \in \mathscr{P}_{\ell-1}, t < \xi_1, t > \xi_k, \xi_i < t < \xi_{i+1}\}$$

are the splines of order ℓ with knots ξ_j and $s \in S_{\ell,k}(\xi)$

$\Leftrightarrow$

$$s(t) = \sum_{i=0}^{\ell-1} u_i t^i + \sum_{i=1}^{k} v_i(t - \xi_i)_+^{\ell-1} .$$

Then we have the following definition: a monospline, m(t), of order ℓ with knots ξ is a function of the form

$$m(t) = \frac{t^\ell}{\ell!} + s(t) , s \in S_{\ell,k}(\xi) .$$

The set of such monosplines is denoted by $M(S_{\ell,k}(\xi))$. We return now to our problem.

Suppose that for some quadrature formula, Q, we have $Rf = 0$, $f \in \mathscr{P}_{r-1}$, then the well-known Peano form of Rf (a simple consequence of the Taylor expansion, with remainder, of f) is

$$\text{(9.2)} \qquad Rf = \int_0^1 f^{(r)}(\zeta)K(\zeta)d\zeta \,,$$

where

$$\text{(9.3)} \qquad K(\zeta) = R \frac{(\cdot - \zeta)_+^{r-1}}{(r-1)!} \,.$$

Substituting (9.3) in (9.2) yields

$$Rf = \int_0^1 f^{(r)}(\zeta)\,[(-1)^r \frac{\zeta^r}{r!} + s(\zeta)]d\zeta \,,$$

for some $s \in S_{r,n}(x)$, and hence

$$\text{(9.4)} \qquad Rf = (-1)^r \int_0^1 f^{(r)}(\zeta)m(\zeta)d\zeta$$

where

$$m \in \overline{M}(S_{r,n}(x)) = \{m \in M(S_{r,n}(x)) : \ m^{(j)}(0) = m^{(j)}(1) = 0, \, j = 0,\ldots,r-1\} \,.$$

Conversely, if $m \in \overline{M}(S_{r,n}(x))$ has the form

$$m(t) = \frac{t^r}{r!} + p_{r-1}(t) + \sum_{i=1}^{n} c_i(t - x_i)_+^{r-1} \,,$$

then integration by parts of (9.4) yields

$$\text{(9.5)} \qquad \int_0^1 f(t)dt = Qf + Rf$$

where

$$\text{(9.6)} \qquad Qf = \sum_{i=1}^{n} -(r-1)!\, c_i f(x_i)$$

and Rf is given by (9.4). Moreover, the formula (9.5) is unique. This is the aforementioned connection (one-to-one relationship) between quadrature formulae and monosplines.

It is easy to see that in our quest for a best quadrature formula in $B_p^r\,[0,1]$ we may assume, with no loss of generality, that $Rf = 0$ for $f \in \mathcal{P}_{r-1}$ and so (9.4) holds. Hölder's inequality now yields

$$\left|\int_0^1 f^{(r)} m\right| \le \left(\int_0^1 |f^{(r)}|^p\right)^{\frac{1}{p}} \left(\int_0^1 |m|^q\right)^{\frac{1}{q}}$$

$$\le \left(\int_0^1 |m|^q\right)^{\frac{1}{q}} = \|m\|_q , \quad \frac{1}{p} + \frac{1}{q} = 1 .$$

Thus

$$\sup_{f \in K} |Rf| = \|m\|_q ,$$

$E(K,x) = \inf \{ \|m\|_q : m \in \overline{M} (S_{r,n}(x))\}$ and

$$i_n(K) = \inf \{ \|m\|_q : m \in \overline{M} (S_{r,n}(x)) , \text{ all nodes } x\} .$$

In plain English, then, the error of a best quadrature formula is the minimum of the L^q norm of allowable monosplines with variable knots, and a minimizing monospline having been found the best quadrature formula is gotten from (9.6). The study of this extremal problem and variants of it obtained by allowing derivative information, or fixing some nodes, or integrating with respect to a weight function, has resulted in an extensive body of literature, which cannot be covered in these lectures. However, we mention that it is known that for all r, n, q there is a unique minimum to this problem (even for integrals with respect to some weight function). See the forthcoming paper of Braess and Dyn [6], Bojanov [5], and references in those works. We look at some examples.

Example 9.1 $r = 1$, $p = \infty$. (Krylov [18])

The allowable monosplines have the form

$$m(t) = t - \sum_{i=1}^{n} a_i(t - x_i)_+^0 , \quad a_1 + \dots + a_n = 1$$

(the case $n = 4$ is illustrated in Fig. 9.1) and

(9.7) $$\int_0^1 |m(t)| \, dt$$

is to be minimized over all allowable choices of a and x.

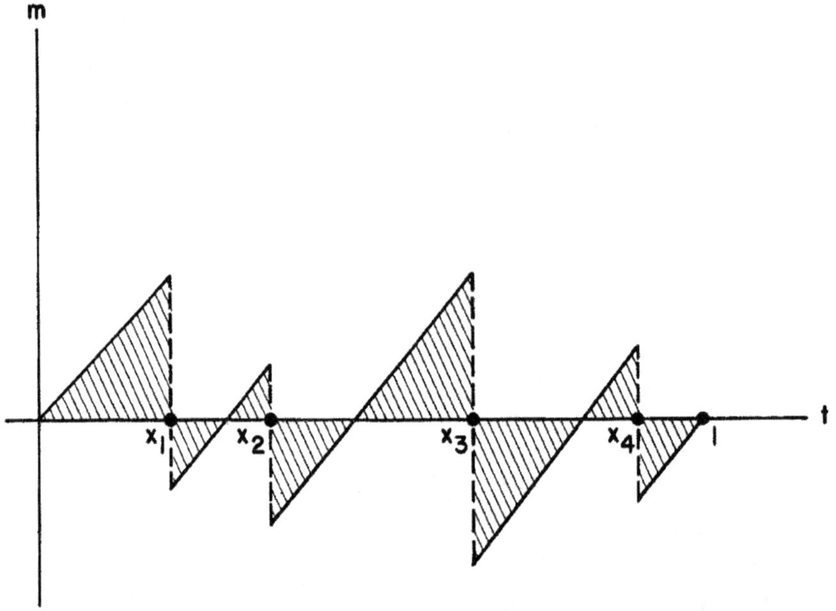

Fig. 9.1

It is clear that (9.7) is minimized when all the hatched triangles have equal area. This is when

$$x_i = \frac{2i-1}{n} \text{ and } a_i = \frac{1}{n}, \quad i = 1,\dots,n .$$

The resulting intrinsic error is $1/(4n)$. The same results hold for $r = 1$, $p \geq 1$.

Example 9.2. In Krylov [18] we also find the result for $r = 2$, $p = \infty$. It is shown there, using Lagrange multipliers, that if $h = (\sqrt{3} + 2(n-1))^{-1}$ the extremal nodes are

$$x_i = \frac{\sqrt{3} + 4(i-1)}{2} h, \quad i = 1,\dots,n$$

and the extremal weights are

$$a_1 = a_n = \frac{2+\sqrt{3}}{2} h \; ; \quad a_i = 2h, \quad i = 2,\dots,n-1 .$$

The resulting intrinsic error is h^2/q . For general r, p the problem seems difficult.

We turn next to the case of functions of period one. $X = \tilde{W}_p^r\,[0,1]$, $K = \tilde{B}_p^r\,[0,1]$ are the corresponding sets of such functions. Zhensykbaev [55] has shown that for all r, p the trapezoidal rule

$$Qf = \frac{1}{n}\sum_{i=1}^{n} f\left(\frac{i}{n}\right) \tag{9.8}$$

is the best quadrature formula. The most difficult part of this result is that the equally spaced points i/n, i = 0,...,n are optimal. However, it has been known for some time that among all quadrature formulae using these nodes the trapezoidal rule is optimal in $\tilde{B}_p^r\,[0,1]$, for all r, p. As the proof of this part of Zhensykbaev's result is elementary and the method applicable in other problems involving optimal quadrature we include it here.

By an argument analogous to that we gave in the non-periodic case we obtain

$$E(\tilde{B}_p^r\,[0,1], x) = \inf\,\{\,\|m\|_q : \ m \in \tilde{M}\,(S_{r,n}(x)\}$$

where $x = \{i/n : \ i = 0,\dots,n\}$ and

$$\tilde{M}\,(S_{r,n}(x)) = \{m \in M\,(S_{r,n}(x)) : \ m^{(j)}(0) = m^{(j)}(1) = 0,\ j = 1,\dots,r-2\}\,.$$

Every $m \in \tilde{M}$ can be extended to a one-periodic function on $\mathbb{R}$ with knots at j/n , j = 0, ±1, ±2,... . Define

$$v(t) = \frac{1}{n+1}\sum_{i=0}^{n} m\left(t + \frac{i}{n}\right),$$

then $v \in \tilde{M}$ has period 1/n, i.e., $v(t + 1/n) = v(t)$. Moreover, $\|v\|_q \le \|m\|_q$, and so the minimum is attained (uniquely for $1 < q < \infty$) by a (1/n)-periodic function. But it is easy to see that the only (1/n)-periodic monospline has the form

$$m(t) = a_0 + \sum_{i=1}^{n} a_i\, D_r(t - x_i)$$

where

$$D_r(t) = 2^{1-r}\pi^{-r}\sum_{j=1}^{\infty} \frac{\cos\left(2\pi jt - \frac{\pi r}{2}\right)}{j^r}\,,$$

the Bernoulli monospline. This same analysis can be used to show the optimality of the Euler-Maclaurin formula for fixed nodes in $B_p^r\,[0,1]$ (see Micchelli [31]).

As for the intrinsic error, if we put

$$\mathcal{K}_{r,p} = \inf_c \| D_r - c \|_p$$

then

$$i_n(K) = \frac{\mathcal{K}_{r,q}}{n^r}\ , \qquad \frac{1}{p} + \frac{1}{q} = 1\,.$$

Let

$$\mathcal{K}_{\ell,1} = \mathcal{K}_\ell = \frac{4}{\pi} \sum_{j=0}^{\infty} \frac{(-1)^{j(\ell+1)}}{(2j+1)^{\ell+1}}$$

be the familiar Favard constants, which satisfy $\mathcal{K}_\ell \le \mathcal{K}_1 = \pi/2$, ℓ odd and $\mathcal{K}_\ell \le 4/\pi$, ℓ even. Then for $p = \infty$

$$i_n(K) = \frac{\mathcal{K}_r}{(2\pi n)^r}\ .$$

Finally, we remark that (9.8) remains best even if we additionally allow first derivatives at x to be sampled. That is, derivative values may be discarded without effect on $i_n(K)$.

9.2 Numerical Integration of Analytic Functions

Let $D = \{z : |z| < 1\}$, $X = H^p$, $1 \le p \le \infty$; $K = \{f \in H^p : \|f\|_p \le 1\}$. Suppose $w : (w_1,\ldots,w_n)$ are given points of D, $Y = \mathbb{C}^n$, If $= (f(w_1),\ldots,f(w_n))$, $\varepsilon = 0$ and

$$Uf = \int_{-1}^{1} f(x)dx\ .$$

Then we know that the intrinsic error E(K ; p,w) satisfies

$$\begin{aligned} E(K) &= \sup\left\{ \left| \int_{-1}^{1} f \right| : \|f\|_p \le 1\ ,\ f(w_i) = 0,\ i = 1,\ldots,n \right\} \\ &= \sup_{\|f\|_p \le 1} \left| \int_{-1}^{1} f(x)B(x)dx \right| \end{aligned}$$

the last equality following from the observation that if

$$B(z) = B_n(z) = \prod_{i=1}^{n} \frac{z - w_i}{1 - \overline{w}_i z}$$

then $f = g\,B \Rightarrow \|g\|_p \le 1$.

The intrinsic error of a best quadrature formula is given by

$$i_n(K,p) = \inf_{B} \sup_{\|f\|_p \le 1} \left| \int_{-1}^{1} f(x)B(x)dx \right| . \tag{9.9}$$

Bojanov [4] shows that when $p = \infty$, **K** is restricted to real-valued functions on $(-1, 1)$ and we deal with 2n real sample points $w_1, w_1, \ldots, w_n, w_n$, so that

$$Qf = \sum_{j=1}^{n} [a_j f(w_j) + b_j\, f'(w_j)] ,$$

we have

$$E(\infty, w) = \int_{-1}^{1} B_n^2(x)dx .$$

It is also shown that an optimal algorithm is given as follows: Put

$$W_k(x) = \frac{x - w_k}{1 - w_k x} \quad ; \quad \omega_k(x) = \prod_{\substack{j=1 \\ i \ne k}}^{n} \frac{x - w_i}{1 - w_i x} ,$$

then

$$b_j = \int_{-1}^{1} \frac{\omega_j^2(x)}{\omega_j^2(x_j)} \, \frac{W_j(x)}{W'_j(x)} \, (1 - W_j^2(x))dx ,$$

and

$$a_j = \int_{-1}^{1} \frac{w_j^2(x)}{w_j^2(x_j)} \, [1 - W_j^4(x) - \frac{2\omega'_j(x_j)W_j(x)}{\omega_j(x_j)W'_j(x_j)} \, (1 - W_j^2(x))]dx, \; j = 1,\ldots,n.$$

Loeb [23] shows that a best formula exists. For such a formula $b_j = 0$, $j = 1,\ldots,n$ and

$$e^{-5\pi\sqrt{\frac{n}{2}}} \le i_{2n}(K, \infty) \le e^{-\pi\sqrt{\frac{n}{2}}} .$$

We return now to the full generality of our original setting. Newman [42] showed that

$$(9.10)\qquad \frac{2}{3}\, e^{-6\sqrt{\frac{n}{q}}} \le i_n(K,p) \le 11e^{-(1/2)\sqrt{\frac{n}{q}}}\ ,\quad \frac{1}{p}+\frac{1}{q}=1\ .$$

We present a sketch of his proof.

(a) Upper bound.

We first recall a result of Szegö (cf. Duren [11]). For $f \in H^p$

$$\left[\int_{-1}^{1} |f(x)|^p dx\right]^{\frac{1}{p}} \le \pi^{\frac{1}{p}} \|f\|_p\ .$$

Thus by Hölder's inequality we have, for any B,

$$\left|\int_{-1}^{1} fB\right| \le \left[\int_{-1}^{1} |f|^p\right]^{\frac{1}{p}} \left[\int_{-1}^{1} |B|^q\right]^{\frac{1}{q}}$$

$$\le \pi^{\frac{1}{p}} \left[\int_{-1}^{1} |B|^q\right]^{\frac{1}{q}}\ .$$

Newman now exhibits an explicit B, using a modification of his famous rational approximation to $|x|$ on $[-1, 1]$ (Newman [41]) which gives the desired upper bound, in view of (9.9).

(b) Lower bound.

For each B consider

$$f_B(z) = \frac{c\overline{B(\overline{z})}}{1-\rho z^2}\ ,\quad \rho = 1 - e^{\sqrt{nq}}\ ,$$

with c chosen so that $\|f_B\|_p = 1$. Then some virtuoso bounding concludes the proof.

The good nodes constructed in (a) are, roughly, at $1 - e^{-\sqrt{n}}$ $1-(e^{-\sqrt{n}} + e^{-\sqrt{n}}/\sqrt{n})$, $1-(e^{-\sqrt{n}} + 2e^{-\sqrt{n}}/\sqrt{n})$,..., which crowd toward the end-points of the interval much faster than the Gauss quadrature nodes, for example. These latter are, roughly, at $1 - n^{-2}$, $1-4n^{-2}$, $1 - 9n^{-2}$,... .

Finally, we mention a recent sharpening of (9.10) due to Andersson [1],

$$\lim_{n\to\infty} [i_n(K, p)]^{\frac{1}{\sqrt{n}}} = e^{-\frac{\pi}{\sqrt{q}}} .$$

10. Miscellaneous.

This final section is devoted to two unrelated problems. The first is optimal numerical differentiation of analytic functions and the second is optimal recovery of best approximations.

10.1 Optimal Numerical Differentiation of Analytic Functions

We are again in the setting given at the beginning of Section 8, except that we seek to recover $f'(\zeta)$ rather than $f(\zeta)$. That is, $X = H^\infty$, $K = \{f \in H^\infty : \|f\| \le 1\}$, and $\zeta, z_1,\dots,z_n \in D$ are given. $Z = \mathbb{C}$, $\varepsilon = 0$, $If = (f(z_1),\dots,f(z_n))$ and $Uf = f'(\zeta)$. Following M-R we divide the unit disc, D, by means of the curve

$$\gamma(\zeta) := \frac{1-|\zeta|^2}{2} \left|\frac{B'(\zeta)}{B(\zeta)}\right| = 1 ,$$

(where

$$B(z) = \prod_{i=1}^{n} \frac{z-z_i}{1-\bar{z}_i z})$$

into two disjoint subsets $D_1 = \{\zeta \in D : \gamma(\zeta) \le 1\}$ and $D_2 = \{\zeta \in D : \gamma(\zeta) > 1\}$. An optimal algorithm is now given by

$$f(z_1,\dots,z_n) \to \sum_{j=1}^{n} c_j(\zeta) f(z_j)$$

where the c_j are determined by

$$f(\zeta) - \sum_{j=1}^{n} c_j\, f(z_j) = \frac{1}{2\pi i} \sum_{|z|=1} \mathcal{K}(x,\zeta) f(z)dz .$$

The kernel, $\mathcal{K}$, has different forms for $\zeta \in D_1$ and $\zeta \in D_2$. Details are in M-R. The two forms of the corresponding intrinsic error are given by

$$E(K\ ;\ z_1,\dots,z_n) = \frac{|B(\zeta)|}{1-|\zeta|^2}\ H(\gamma(\zeta)) \tag{10.1}$$

where

$$H(\gamma) = \begin{cases} 1+\gamma^2\ , & 0 \le \gamma \le 1\ , \\ 2\gamma\ , & \gamma > 1\ . \end{cases}$$

Note that H is continuously differentiable on $\mathbb{R}^+$.

We wish to determine the optimal sampling points in a rather simple situation, as described in Rivlin, Ruscheweyh, Shaffer and Wirths [48]. Suppose $-1 < z_1 \le z_2 \le \dots \le z_n \le 0$ and $0 < \zeta < 1$. Our problem is to minimize the intrinsic error, as given in (10.1), with respect to $z_1,\dots,z_n$. To this end let us put

$$b_j = \frac{\zeta - z_j}{1 - z_j\zeta}\ ,\quad j = 1,\dots,n,$$

then the b_j and z_j are in one-to-one correspondence, and b_j varies over $[\zeta,1)$ as z_j varies over $(-1,0]$. Note that now

$$\gamma(\zeta) = \frac{1}{2}\sum_{j=1}^{n} \frac{1-b_j^2}{b_j}\ .$$

Hence the intrinsic error satisfies

$$(1-\zeta^2)\ E\ (K\ ;\ z_1,\dots,z_n) = (\prod_{j=1}^{n} b_j)\ H\ (\frac{1}{2}\sum_{j=1}^{n} \frac{1-b_j^2}{b_j})\ . \tag{10.2}$$

We shall determine the minimum of $E(K\ ;\ z_1\dots z_n)$ for $(b_1,\dots,b_n) \in [\zeta,1)^n$, and where it is attained. We restrict our attention to $n \ge 2$. The case of $n = 1$ is simple, but requires a slightly different analysis. We shall treat it later.

The function $g(y) = -n \sinh y$ is convex for $y < 0$. If we put $y_j = \log b_j$, $j = 1,\dots,n$ and

$$y_0 = \frac{1}{n}\sum_{j=1}^{n} y_j\ ,$$

then

$$g(y_0) \leq \frac{1}{n} \sum_{j=1}^{n} g(y_j) ,$$

or, equivalently,

$$\frac{n}{2} \frac{1-b^2}{b} \leq \frac{1}{2} \sum_{j=1}^{n} \frac{1-b_j^2}{b_j} , \tag{10.3}$$

where

$$b = e^{y_0} = \prod_{j=1}^{n} b_j^{1/n} \in [\zeta, 1) .$$

Since H is a monotone increasing function (10.2) and (10.3) yield

$$(1-\zeta^2)\, E\,(K\, ;\; z_1,...,z_n) \geq b^n\, H\,(\frac{n}{2}\, \frac{1-b^2}{b}) \;=:\, F_n(b) . \tag{10.4}$$

It is not difficult to see that the function $F_n(t)$, $t \in (0, 1]$, has exactly two local extrema in (0,1), a maximum at $t = M(n) = ((n-1)/(n+1))^{1/2}$ and a minimum at $t = \tau_1(n) = (n/(n+2))^{1/2}$. In the intervals of (0,1) which exclude these points F_n is strictly monotone. Since

$$\lim_{t \to 0} F_n(t) = 0$$

it is clear that there exists exactly one solution of the equation $F_n(t) = F_n(\tau_1)$ in $(0, \tau_1)$.

Call this solution $\tau_0(n)$. A typical graph of $F_n(t)$ is shown in Figure 10.1.

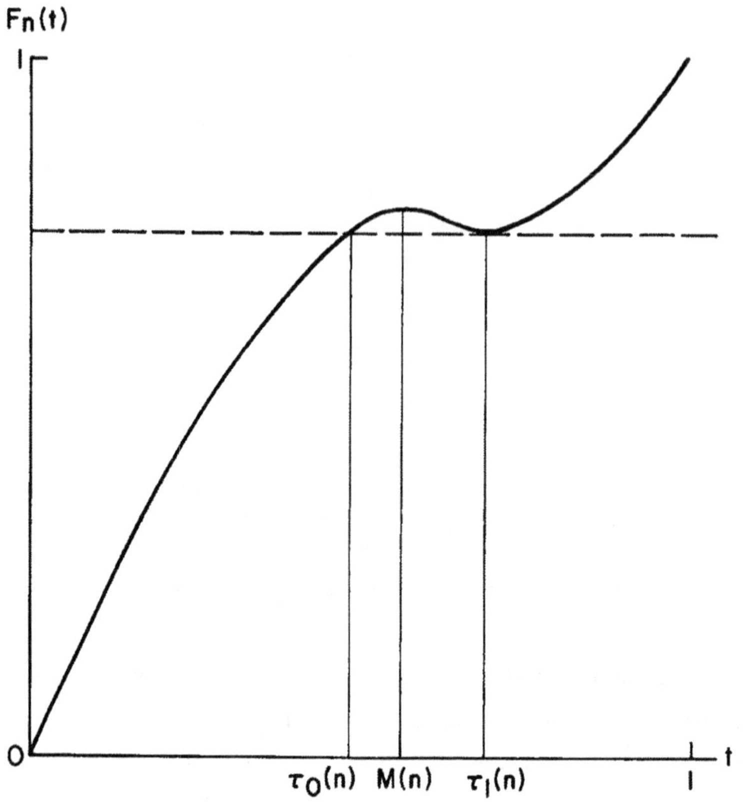

Fig. 10.1

Now

(10.5) $$F_n(b) \geq \min\{F_n(t) : t \in [\zeta,1)\} =: (1-\zeta^2)\, G_n(\zeta) .$$

But our discussion of the behavior of $F_n(t)$ yields

(10.6) $$G_n(\zeta) = \begin{cases} n\,\zeta^{n-1}, & 0 < \zeta \leq \tau_0 \\ n\tau_0^{n-1}\,\dfrac{1-\tau_0^2}{1-\zeta^2}, & \tau_0 < \zeta \leq \tau_1 \\ \dfrac{\zeta^n}{1-\zeta^2}\,[1 + \dfrac{n^2}{4}\,(\dfrac{1-\zeta^2}{\zeta})^2], & \tau_1 < \zeta < 1 , \end{cases}$$

if we observe that the positive zero of $t^2 + 2n^{-1}t - 1$ is greater than $M(n)(> \tau_0)$. Thus if we put

(10.7) $$z(\zeta) = \begin{cases} \dfrac{\zeta-\tau_1}{1-\tau_1\zeta}, & \tau_0 < \zeta < \tau_1 \\ 0, & \text{elsewhere in } (0,1), \end{cases}$$

we conclude from (10.4) - (10.7) that $E(K\,;\ z_1,\dots,z_n)$ attains its minimum for $z_1 = z_2 = \dots = z_n = z(\zeta)$ and this minimum is $G_n(\zeta)$. The resulting optimal algorithm for the optimal sampling $(f(z(\zeta)), f'(z(\zeta)),\dots,f^{(n-1)}(z(\zeta))$ is fully described in Rivlin, Ruscheweyh, Shaffer and Wirths [48].

We remark that it can be shown that $\tau_0(n) > ((n-2)/n)^{1/2}$. This inequality informs us that, for large n, $(0, \tau_0)$ covers most of the range of ζ. Also note that for $0 < \zeta < \tau_0$, $z(\zeta) = 0$. This leads to the striking fact that, for $0 < \zeta < \tau_0$, the optimal algorithm is simply the derivative with respect to ζ of the optimal algorithm for *interpolation* at ζ in the case $z_1 = z_2 = \dots = z_n = 0$ as given by (8.2).

Finally, we turn to the case n = 1, as promised. The discussion proceeds as in the previous case except that the function $F_1(t)$ is degenerate. Namely, if $z_1 = \lambda$ and

$$b = \frac{\zeta-\lambda}{1-\lambda\zeta},$$

$(1-\zeta^2)E(K;\ \lambda) = bH(\frac{1}{2}\ \frac{1-b^2}{b}) = F_1(b)$. Now $F_1(t)$, $t \in (0,1]$ has exactly one local extremum in (0,1), a minimum at $1/3^{1/2}$, and $F_1(0) = F_1(1) = 1$. The result is that $E(K, \lambda)$ attains its minimum for

$$\lambda(\zeta) = \begin{cases} \dfrac{\sqrt{3}\,\zeta-1}{\sqrt{3}-\zeta}, & 0 \le \zeta < 3^{-1/2}, \\ 0, & 3^{-1/2} \le \zeta < 1, \end{cases}$$

and the minimum is given by

$$E(K) = \begin{cases} \dfrac{4\sqrt{3}}{9}\ \dfrac{1}{1-\zeta^2}, & 0 \le \zeta < 3^{-1/2}, \\ \dfrac{(1+\zeta^2)^2}{4\zeta(1-\zeta^2)}, & 3^{-1/2} \le \zeta < 1. \end{cases}$$

Note the surprising fact that in order to recover $f'(0)$ it is best to back away to $\lambda(0) = -1/\sqrt{3}$. Full details are in Rivlin and Shaffer [49].

10.2 Optimal Recovery of Best Approximations

If $f \in C[-1,1]$ there exists a polynomial of degree at most n, $p_n(f)$, having the property that

$$\|f - p_n(f)\| < \|f - p\| , \text{ all } p \in \mathcal{P}_n , p \neq p_n(f) .$$

($\|\bullet\|$ is the maximum norm on $[-1,1]$). The problem we wish to consider is that of the optimal recovery of $p_n(f)$ for f in the unit ball in $C[-1,1]$ from a fixed finite number of sampled values of f. In our usual formulation, then, we have $X = Z = C[-1,1]$, $K = \{f \in X : \|f\| \leq 1\}$. Let $a = (a_1,...,a_N)$, $-1 \leq a_1 < a_2 <...< a_N \leq 1$, be given. Suppose $If = (f(a_1),...,f(a_N))$, $Y = \mathbb{R}^N$ (with the maximum norm), $\varepsilon \geq 0$ and $Uf = p_n(f)$. Apart from its intrinsic interest this is an example of a problem with a non-linear feature operator, Uf, a situation we have not previously encountered. Observe that $U(-f) = -Uf$ and so (2.1) is valid, as we remarked in the note following Theorem 2.1. The solution we shall give comes from Micchelli and Rivlin [38].

To help solve this problem a solution of the following polynomial extremal problem will be helpful. Let

$$L(x) = \sup \{ |p(x)| : p \in \mathcal{P}_n , |p(a_i)| \leq 1 , i = 1,...,N\} , \tag{10.8}$$

and

$$\Lambda = \sup \{L(x) : -1 \leq x \leq 1\} . \tag{10.9}$$

When $n \geq 1$, which we assume henceforth, the determination of L(x) can be characterized as follows. Suppose $N \geq n + 1$. Let $J \subset \{1,...,N\}$ consist of any n+1 distinct integers. Put $a(j) = \{a_j : j \in J\}$. $\ell_i(a(J),x)$, $i = 1,...,n+1$ are the fundamental polynomials for interpolation on a(J), and

$$\lambda_n(a(J),x) = \sum_{i=1}^{n+1} |\ell_i(a(J),x)|$$

is the Lebesgue function for a(J). Micchelli and Rivlin [38] show that for $N \geq n + 1$

$$L(x) = \min_J \lambda_n(a(J),x) , \tag{10.10}$$

and it is easy to see that for $N \leq n$

$$(10.11) \qquad L(x) = \begin{cases} \infty, & x \notin a \\ 1, & x \in a . \end{cases}$$

Moreover, if the minimum in (10.10) is attained for some J for an $x \in (a_k, a_{k+1})$ then the minimum is attained for the same J for all $x \in (a_k, a_{k+1})$. It can now be shown that

$$(10.12) \; e(K,\varepsilon) = \sup\{ \| p_n(f) \| : f \in K, \ |f(a_i)| \leq \varepsilon, i = 1,\ldots,N\} \geq \min (2,(1 + \varepsilon)\Lambda).$$

With these facts established an optimal algorithm may be determined. There are two cases to consider.

(i) $(1+ \varepsilon)\Lambda < 2$.

The algorithm

$$\alpha : \ y = (y_1,\ldots,y_n) \to \sum_{i \in J(x)} y_i \, \ell_i(a(J(x)), x) ,$$

is optimal, where, given x, J(x) is the subset for which the minimum in (10.10) is attained. Note that α produces a piecewise polynomial approximation to $p_n(f)$.

(ii) $2 \leq (1 + \varepsilon)\Lambda$.

The algorithm $\alpha' : \ y \to 0$ is optimal.

Let us prove that α and α', respectively, are optimal. In case (i), since $N \leq n$ implies that $\Lambda = \infty$, we must have $N \geq n + 1$. Then

$$\| p_n(f) - \alpha y \| = \max_{-1 \leq x \leq 1} \Big| \sum_{i \in J(x)} [p_n(f)(a_i) - y_i] \, \ell_i(a(J(x)), x) \Big| .$$

But since $\| p_n(f) - f \| \leq 1$ and $| f(a_i) - y_i | \leq \varepsilon$ the triangle inequality yields

$$\| p_n(f) - \alpha y \| \leq (1 + \varepsilon) \max_{-1 \leq x \leq 1} L(x) = (1 + \varepsilon)\Lambda \leq e(K,\varepsilon) ,$$

and α is optimal. The intrinsic error is $(1 + \varepsilon)\Lambda$. In case (ii), $\| f - p_n(f) \| \leq \| f \| \Rightarrow \| p_n(f) \| \leq 2 = e(K,\varepsilon)$ and α' is optimal. The intrinsic error is 2.

The case $n = 0$, which we did not consider, is particularly simple. The best approximating constant to an $f \in K$ is

$$p_0(f) = \frac{\max_{-1 \leq x \leq 1} f(x) + \min_{-1 \leq x \leq 1} f(x)}{2} .$$

Thus $\|p_0(f)\| \leq 1$ and it is easy to see that $e(K,\varepsilon) = 1$. But then α' , as defined in (ii) above, is optimal.

Finally, we remark that we can give a rough asymptotic bound for the intrinsic error $(n > 1)$, as $N \to \infty$, in some cases. Suppose $a_1 = -1$, $a_N = 1$ and

$$\Delta = \max_i (a_{i+1} - a_i)$$

satisfies

$$\Delta^2 < \frac{12}{n^2(n^2-1)} ,$$

then

$$\Lambda \leq 1 + \frac{\Delta^2}{12} n^2(n^2-1) .$$

For details see Micchelli and Rivlin [38].

Papers and books referred to in the preceding text are listed next. In addition we wish to direct the reader's attention to the pioneering paper of Golomb and Weinberger [17], which is still well worth reading, the book of Traub and Wozniakowski [52] which has a point of view similar to ours and contains an annotated bibliography of over 300 items, which is particularly valuable for its coverage of Soviet and other Eastern European literature, and to the selected survey by Rivlin [47].

References

1. Andersson, J-E., Optimal quadrature of H^p functions, Math. Z. **172**, 1980, 55-62.

2. Babadjanov, S. B. and V. M. Tichomirov, On the width of a functional class in the space L_p $(p \geq 1)$, Izv. Akad. Nauk UZSR, Ser. Fiz-Mat. Nauk **12**, 1967, 24-30.

3. Berger, Melvin, and Marion Berger, "Perspectives in Nonlinearity", W. A. Benjamin, Inc., N. Y., 1968.

4. Bojanov, B. D., Best quadrature formula for a certain class of analytic functions, Zastos. Mat., XIV, 1974, 441-447.

5. Bojanov, B. D., Uniqueness of monosplines of least deviation, "Numerische Integration", Ed. G. Hämmerlin, International Series of Numerical Mathematics, Vol. 45, Birkhäuser, Basel, 1979, 67-97.

6. Braess, D., and N. Dyn, On the uniqueness of generalized monosplines with least L_p norm, to appear.

7. Cheney, E. W., "Introduction to Approximation Theory", McGraw-Hill Book Co., N. Y., 1966.

8. Dahmen, W., Micchelli, C. A. and P. W. Smith, Asymptotically optimal sampling schemes for periodic functions, IBM Research Report, RC 10733, 1984.

9. Davis, C., Kahan, W. M., and H. F. Weinberger, Norm-preserving dilations and their applications to optimal error bounds, SIAM J. Numer. Anal. **19**, 1982, 445-469.

10. Duchon, J., Splines minimizing rotation-invariant semi-norms in Sobolev spaces, "Constructive Theory of Functions of Several Variables", Eds. W. Schempp and K. Zeller, Springer, Berlin -Heidelberg, 1976, 85-100.

11. Duren, P., "Theory of H^p Spaces", Academic Press, N. Y., 1970.

12. Fisher, S. D., and C. A. Micchelli, The n-width of sets of analytic functions, Duke Math. J. **47**, 1980, 789-801.

13. Fisher, S. D., and C. A. Micchelli, Optimal sampling of holomorphic functions, Amer. J. Math. **106**, 1984, 593-609.

14. Gaffney, P. W., and M.J.D. Powell, Optimal interpolation, "Numerical Analysis" (Proc. 6th Biennial Dundee Conf., Univ. Dundee, 1975). Lecture Notes in Math., Vol. 506, Springer, Berlin, 1976, 90-99.

15. Gal, S., and C. A. Micchelli, Optimal sequential and non-sequential procedures for evaluating a functional, Applicable Analysis **10**, 1980, 105-120.

16. Gantmacher, F. R., and M. G. Krein, "Oscillationsmatrizen, Oscillationskerne und kleine Schwingungen mechanischer Systeme", Akademie-Verlag, Berlin, 1960.

17. Golomb, M. and Weinbeiger, H. F., Optimal approximation and error bounds, "On Numerical Approximation", Ed. R. Langer, U. of Wisconsin Press, Madison, Wisconsin, 1959, 117-190.

17′ Höllig, K., A generalization of Jackson's inequality, J. Approx. Theory **31**, 1981, 154-157.

18. Krylov, V. I., "Approximate Calculation of Integrals", Macmillan, N. Y. 1962.

19. Kuo, H.-H., "Gaussian Measures in Banach Spaces", Lecture Notes in Mathematics, Vol. 463, Springer, Berlin, 1975.

20. Larkin, F. M., Gaussian measure in Hilbert space and application to numerical analysis, Rocky Mt. J. Math. **2**, 1972, 379-421.

21. Laüter, H., A minmax linear estimator for linear parameters under restrictions in form of inequalities, Math. Operationsforsch. Statist. **6**, 1975, 689-695.

22. Levin, M., and J. Girshovich, "Optimal Quadrature Formulas", B. G. Teubner, Leipzig, 1979.

23. Loeb, H. L., A note on optimal integration in H_∞, C. R. Akad. Bulgare Sci. **27**, 1974, 615-618.

24. Logan, B. F., and L. A. Shepp, Optimal reconstruction of a function from its projections, Duke Math. J. **42**, 1975, 645-659.

25. Makovoz, Yu., I., On a method for estimation from below of diameters of sets in Banach spaces, Math. U.S.S.R. Sb. **16,** 1972, 139-146.

26. Meinguet, J., An intrinsic approach to multivariate spline interpolation at arbitrary points, "Polynomial and Spline Approximation", Ed., B. N. Sahney, D. Reidel, Dordrecht, 1979, 163-190.

27. Melkman, A. A., n-widths and optimal interpolation of time-and band-limited functions, "Optimal Estimation in Approximation Theory", Eds., C. A. Micchelli and T. J. Rivlin, Plenum Press, N. Y., 1977, 55-68.

28. Melkman, A. A. and C. A. Micchelli, Spline spaces are optimal for L^2 n-widths, Illinois J. Math. **22**, 1978, 541-564.

29. Melkman, A. A. and C. A. Micchelli, Optimal estimation of linear operators in Hilbert spaces from inaccurate data, SIAM J. Numer. Anal. **16**, 1979, 87-105.

30. Melkman, A. A., n-widths and optimal interpolation of time- and band-limited functions II, SIAM J. Math. Anal., to appear.

31. Micchelli, C. A., Best quadrature formulas at equally spaced nodes, J. Math. Anal. Appl. **47**, 1974, 232-249.

32. Micchelli, C. A., Optimal estimation of smooth functions from inaccurate data, J. Inst. Maths. Applics. **23**, 1979, 473-495.

33. Micchelli, C. A., On Weinberger's duality formula for the approximation of bounded operators on Hilbert spaces, IBM Research Report, RC #7722, 1979.

34. Micchelli, C. A., Orthogonal projections are optimal algorithms, J. Approx. Theory **40**, 1984, 101-110.

35. Micchelli, C. A.and A. Pinkus, Some problems on the approximation of functions of two variables and n-widths of integral operators, J. Approx. Theory **24**, 1978, 51-77.

36. Micchelli, C. A., T. J. Rivlin and S. Winograd, The optimal recovery of smooth functions, Numer. Math. **26**, 1976, 191-200.

37. Micchelli, C. A. and T. J. Rivlin, A survey of optimal recovery, "Optimal Estimation in Approximation Theory" (Eds. C. A. Micchelli and T. J. Rivlin), Plenum, N. Y., 1977, 1-54.

38. Micchelli, C. A. and T. J. Rivlin, Optimal recovery of best approximations, Resultate der Math. **3**, 1980, 25-32.

39. Micchelli, C. A., and G. Wahba, Design problems for optimal surface interpolation, "Approximation Theory and Applications", Ed. Z. Ziegler, Academic Press, N. Y., 1981, 329-348.

40. Mirsky, L., Symmetric gauge functions and unitarily invariant norms, Quart. J. Math. Oxford Ser. (2) **11**, 1960, 50-59.

41. Newman, D. J., Rational approximation to $|x|$, Michigan Math. J. **11**, 1964, 11-14.

42. Newman, D. J., Quadrature in H^p, Lectures III, IV, "Approximation with Rational Functions", CBMS Regional Conference Series in Math., No. 41, Amer. Math. Soc., Providence, Rhode Island, 1979.

43. Osipenko, K. Yu, Optimal interpolation of analytic functions, Math. Notes Acad. Sci. U.S.S.R. **12**, 1972, 712-719.

44. Osipenko, K. Yu., Best methods of approximating analytic functions given with an error, Math. U.S.S.R. Sbornik **46**, 1983, 353-374.

45. Pinkus, A., n-widths of Sobolev spaces in L^p, to appear in Constructive Approximation.

46. Rivlin, T. J., A survey of recent results on optimal recovery, "Polynomial and Spline Interpolation", Ed. B. N. Sahney, D. Reidel, Dordrecht, 1979, 225-245.

47. Rivlin, T. J., The optimal recovery of functions, Contemporary Math., Vol. 9, 1982, 121-151.

48. Rivlin, T. J., St. Ruscheweyh, D. Shaffer and K. J. Wirths, Optimal recovery of the derivative of bounded analytic functions, IMA J. of Numer. Anal. **3**, 1983, 327-332.

49. Rivlin, T. J., and D. B. Shaffer, Optimal estimation of the derivative of bounded analytic functions, IBM Research Report, RC #9843, 1983.

50. Scharlach, R., Optimal recovery by linear functionals, J. Approx. Theory, to appear.

51. Speckman, P., Minmax estimates of linear functionals in a Hilbert space, preprint.

52. Traub, J. F. and H. Wozniakowski, "A General Theory of Optimal Algorithms", Academic Press, N. Y., 1980.

53. Weinberger, H. F., On optimal numerical solution of partial differential equations, SIAM J. Numer. Anal. **9**, 1972, 182-198.

54. Zensykbaev, A. A., Optimal recovery methods for the integral on classes of differentiable functions, Analysis Mathematica **7**,1981, 303-318.

55. Zhensykbaev, A. A., Extremal properties of monosplines and best quadrature formulas, Mathematical Notes **31**, 1982, 145-154.

An Introduction to the Analysis of the Error in The Finite Element Method for Second-Order Elliptic Boundary Value Problems

A.H. Schatz
Department of Mathematics
Cornell University
Ithaca, New York 14853

Introduction

The aim of these notes is to provide an introduction to the mathematical analysis of the finite element method for second order elliptic boundary value problems. No attempt has been made to give an exhaustive or general treatment of the subject. In fact only a few topics will be covered and these treated in certain model situations. An attempt has been made to present some of the basic ideas and techniques used in obtaining error estimates for these problems and some results with selfcontained proofs. It is hoped that this will provide the reader with enough tools to be able to generalize the results given here and also understand some of the more recent developments in the field. An outline of these notes is as follows:

Section 1 contains some definitions and preliminaries concerning the various function spaces which will be used in subsequent sections. For example we shall consider some Sobolev spaces and prove some elementary Poincaré inequalities.

In section 2, two model elliptic boundary value problems are formulated and some properties of their solutions discussed. More precisely we consider the weak formulation of Dirichlet's problem for the Poisson's equation and a model Neumann problem on a bounded domain in the plane. Some apriori estimates for these problems will be stated under various assumptions on the data and smoothness of the boundary. These will include plane polygonal domains.

In Section 3 the finite element spaces which will be used to approximate the solutions of the above boundary value problems are defined. For simplicity we shall discuss in detail the space of continuous piecewise linear functions defined on a quasi-uniform triangulation of size h of a two dimensional domain Ω. Various basic properties of these functions are studied, for example, approximation properties, inverse properties and the superapproximation property. This is first done when the domain Ω is polygonal. Other types of domains and some other finite element spaces are briefly discussed at the end.

In Section 4 the Galerkin method is introduced in the context of one of the model problems. The finite element method is a special case of this method with a special choice of approximating functions. Existence and uniqueness of a solution are discussed and the solution is characterized as an orthogonal projection of the solution into the space of approximating functions. This leads immediately to error estimates in the so called energy norm provided some approximation properties of the approximating functions in this norm are known. The case of the above mentioned piecewise linear functions are then discussed in some detail for problems on polygonal domains. Error estimates are then derived in $L^2(\Omega)$ using a very important duality argument due to Aubin and Nitsche. The case of a non-convex polygonal domain is discussed as an example where "non-optimal rates of convergence" are obtained. Finally error estimates are discussed for domains with smooth boundaries and when some other approximating spaces are used.

In Section 5 we shall discuss local error estimates for the finite element method. More precisely we obtain energy estimates for the error on subdomains of Ω and see to what extent they are influenced by the nature of the solution of the boundary value problem and of the boundary outside of this subdomain. In many important situations, the rate of convergence of the approximating solution to the true solution is higher in certain regions than in others. This for example occurs in our model problems when the domain Ω is a non-convex polygon. Under the appropriate assumptions on the data, convergence will be faster away from non-convex corners than near the corners. Our aim in treating this problem is two fold. First, it is of interest in itself and the results have been used in treating other problems. Secondly, we will apply these results directly to the problems of finding maximum norm error estimates.

In Section 5 maximum norm error estimates for Dirichlet's problem on a plane convex polygonal domain are derived. There are several methods which have been developed which could be used to prove the result given here. Two of them will be outlined and a third given in detail. All three have something in common in that their proofs use some of the ideas used in the proofs of local estimates given in chapter 5.

1. Notation and Some Preliminaries

Throughout these notes the notation of Showalter [16] will be used. In addition, for the bounded domains Ω used here (which have either a smooth boundary or are polygonal domains in $\mathbb{R}^2$) $C^m(\bar{\Omega})$; $m \geq 0$ an integer, will denote the space of functions having m continuous derivatives on $\bar{\Omega}$. If $m = 0$ we set $C^0(\Omega) = C(\bar{\Omega})$. This is a Banach space with the norm

$$\|u\|_{C^m(\bar{\Omega})} = \sum_{|\alpha| \leq m} \max_{x\in\bar{\Omega}} |D^\alpha u| .$$

We shall also use semi-norms defined on $C^m(\bar{\Omega})$ and $W^{m,p}(\Omega)$ (distinguised by single bars) defined by

$$|u|_{C^m(\bar{\Omega})} = \sum_{|\alpha| = m} \max_{x\in\bar{\Omega}} |D^\alpha u| ,$$

and

$$|u|_{m,p,\Omega} = \Big(\sum_{|\alpha|=m} \|\partial^\alpha u\|^p_{L^p(\Omega)} \Big)^{1/p}, \quad 1 \leq p < \infty$$

respectively. We set

$$\nabla u = (\partial_1 u; \ldots, \partial_N u)$$

the gradient of the function u and

$$\|\nabla u\|_{L^p(\Omega)} = |u|_{1,p,\Omega} .$$

Furthermore we shall use

$$(u,v) = \int_\Omega uv\, dx .$$

In the following two lemmas we shall prove some useful Poincaré inequalities.

Lemma 1.1. Let Ω be either a square in $\mathbb{R}^2$ with edges of length $d > 0$ or the triangle with vertices (0,0), (d,0), (0,d). Then

(i) if $u \in W^{1,2}(\Omega)$.

$$\|u\|^2_{L^2(\Omega)} \leq d^2 \|\nabla u\|^2_{L^2(\Omega)} + d^{-2} \Big(\int_\Omega u dx\Big)^2 . \tag{1.1}$$

(ii) If $u \in C^1(\bar{\Omega})$,

$$(1.2) \qquad \|u\|_{C(\bar{\Omega})} \leq C(d\|\nabla u\|_{C(\bar{\Omega})} + d^{-2} \,|\int_\Omega u dx|),$$

where C is independent of u and d.

<u>Proof</u>: Since $C^1(\bar{\Omega})$ is dense in $W^{1,2}(\Omega)$ it suffices to consider $u \in C^1(\bar{\Omega})$ in both (1.1) and (1.2). Let us first prove (1.2). Let $x_0 \in \bar{\Omega}$ be such that $\sup_{x\in\bar{\Omega}} |u(x)| = |u(x_0)|$. If $u(x_0) = 0$, (1.2) is trivial. In the case that $u(x_0) > 0$, we have for any $y \in \bar{\Omega}$ and the mean value theorem (since both domains are convex)

$$0 \leq u(x_0) - u(y) \leq \sqrt{2}\, d\, \|\nabla u\|_{C(\bar{\Omega})}$$

integrating with respect to y_1, y_2. We obtain

$$u(x_0) \leq C(d\|\nabla u\|_{C(\bar{\Omega})} + d^{-2} \int_\Omega |u(y)dy)|$$

If $u(x_0) < 0$ then by a similar argument we obtain

$$-u(x_0) \leq C(d\|\nabla u\|_{C(\bar{\Omega})} + d^{-2} \,|\int_\Omega u(y)dy|$$

Taken together these last two inequalities imply (1.2). To prove (1.1) in the case of the square we use the identity

$$(u(x) - u(y))^2 = (\int_{y_1}^{x_1} \frac{\partial u}{\partial \zeta} (\zeta, y_2) d\zeta + \int_{y_2}^{x_2} \frac{\partial u}{\partial \eta} (x_1, \eta) d\eta)^2$$

and the Schwarz inequality to obtain

$$(1.3) \qquad u(x)^2 + u^2(y) \leq 2d^2 (\int_0^d (\frac{\partial u}{\partial \zeta} (\zeta, y_2))^2 \, d\zeta + \int_0^d (\frac{\partial u}{\partial \eta} (x, \eta))^2)$$

$$+ 2\, u(x)\, u(y)\, .$$

Integrating with respect to x_1, x_2, y_1, y_2 we obtain

$$2d^2 \|u\|^2_{L^2(\Omega)} \leq 2d^4 \, \|\nabla u\|_{L^2(\Omega)} + 2(\int_\Omega u \, dx)^2$$

which completes the proof of (1.1) in the case of a square. In the case of triangle we reflect u as an even function across the hypotenuse. The extended function, say u, is continuous on the square and piecewise C^1 on each of the triangles. The inequality (1.3) still holds for u on the square and (1.1) follows easily from the fact that u was extended as an even function.

Lemma 1.2. Let Ω be a bounded open set in $\mathbb{R}^2$, $\Omega \subset D$ where $D = \{(x,y) : 0 < x_1 < d,\ 0 < x_2 < d\}$. Then if $u \in W_0^{1,2}(\Omega)$

$$\|u\|_{L^2(\Omega)} \leq d\|\nabla u\|_{L^2(\Omega)} \tag{1.4}$$

Proof: By density it suffices to consider $u \in C_0^1(\bar{\Omega})$, and extend u by 0 to D. Then since $u(0,x_2) = 0$

$$u(x_1,x_2) = \int_0^{x_1} \frac{\partial}{\partial \zeta} (\zeta,x_2)d\zeta .$$

Squaring and using the Cauchy-Schwarz inequality

$$u^2(x_1,x_2) \leq d \int_0^d (\frac{\partial u}{\partial \zeta} (\zeta,x_2))^2 \, d\zeta .$$

The result now easily follows after integrating both sides with respect to x_1 and x_2.

Remark 1.1. An immediate consequence of (1.4) is that for $u \in W_0^{1,2}(\Omega)$, where Ω is an open bounded set with $\mathrm{diam}(\Omega) = d$,

$$\|\nabla u\|^2_{L^2(\Omega)} \leq \|u\|^2_{1,2,\Omega} \leq (1 + d^2)\|\nabla u\|^2_{L^2(\Omega)}$$

Hence $(\nabla u,\nabla v)$ is an inner produce on $W_0^{1,2}$ with an equivalent norm.

We shall now state a special case of the so called Sobolev imbedding theorems.

Lemma 1.3. Let Ω be a bounded domain in $\mathbb{R}^2$ with a boundary $\partial\Omega$ which is either C^1 or polygonal. If $mp > 2$ then $W^{m,p}(\Omega) \subset C(\bar{\Omega})$ and

$$\|u\|_{C(\bar{\Omega})} \leq C\|u\|_{m,p,\Omega} \tag{1.5}$$

2. Two Model Elliptic Boundary Value Problems.

In this section we shall discuss two model elliptic boundary value problems to which the finite element method will be applied. In order to estimate the error between the approximate solution obtained by this method and the solution, say u, of the boundary value problem, we need some information regarding the smoothness of u. This in general depends on several factors which will be discussed below in various situations.

Let Ω be a bounded domain in $\mathbb{R}^2$ and for say $f \in L^2(\Omega)$ let $u \in W_0^{1,2}(\Omega)$ be the weak solution of Dirichlets' problem

(2.1) $$\begin{aligned} -\Delta u &= f \text{ in } \Omega \\ u &= 0 \text{ on } \partial\Omega. \end{aligned}$$

More precisely u is characterized in the following way:

Problem I. For $f \in L^2(\Omega)$, find $u \in W_0^{1,2}(\Omega)$ satisfying

(2.2) $$(\nabla u, \nabla\phi) = (f,\phi) \quad \forall\ \phi \in W_0^{1,2}(\Omega)$$

It is well known that Problem I has unique solution. In fact this is also true if we replace $f \in L^2(\Omega)$ with $f \in (W_0^{1,2}(\Omega))'$ the dual space of $W_0^{1,2}(\Omega)$ see Showalter [16].

Let $f \in L^2(\Omega)$. We know that $u \in W_0^{1,2}(\Omega)$ and it is reasonable to ask whether u is smoother e.g. does $u \in W^{2,2}(\Omega)$? This depends on the nature of $\partial\Omega$. We shall state some regularity results for some classes of domains which will be considered in the following sections.

Theorem 2.1. Let $u \in W_0^{1,2}(\Omega)$ satisfy (2.2).

(i) If Ω has a C^r, $r \geq 2$, boundary $\partial\Omega$ and $f \in W^{r-2,2}(\Omega)$ then $u \in W^{r,2}(\Omega)$ and

(2.3) $$\|u\|_{r,2,\Omega} \leq C\|f\|_{r-2,2,\Omega}\ .$$

(ii) If Ω is a convex domain (for example a convex polygonal domain) and $f \in L^2(\Omega)$, then $u \in W^{2,2}(\Omega)$ (2.3) holds with $r = 2$.

(iii) If Ω is a non-convex polygonal domain with interior angles $0 < \alpha_1 \leq \alpha_2 \leq \ldots\ldots < \alpha_m \leq 2\pi$, $\pi < \alpha_m \leq 2\pi$, and $f \in L^2(\Omega)$, (or even $C^\infty(\bar{\Omega})$), then in general $u \in W^{2,p-\varepsilon}(\Omega)$ for every $\varepsilon > 0$ where $p = \dfrac{2\alpha_m}{2\alpha_m - \pi} < 2$, and in general $u \notin W^{2,p}(\Omega)$. If $f \in L^p(\Omega)$ then

$$(2.4) \qquad \|u\|_{2,p-\varepsilon,\Omega} \leq C_\varepsilon \|f\|_{L^p(\Omega)} \quad \text{for any} \quad \varepsilon > 0.$$

The proof of Theorem 2.1 may be found for example in Grisvard [6].

We shall also be concerned with weak solutions of the Neumann problem

$$(2.5) \qquad -\Delta u + au = f \text{ in } \Omega$$

$$\frac{\partial u}{\partial n} = 0 \qquad \text{on } \partial\Omega.$$

where $a > 0$ is a constant and $\frac{\partial u}{\partial n}$ denote the outward drawn normal derivative to $\partial\Omega$. More precisely u is characterized as follows.

Problem II. For $f \varepsilon (W^{1,2}(\Omega))'$, and $u \varepsilon W^{1,2}(\Omega)$ satisfying

$$(2.6) \qquad (\nabla u, \nabla\phi) + (au, \phi) = (f, \phi).$$

The existence and uniqueness of a solution of Problem II is well known. Furthermore we have (Grisvard [6]).

Theorem 2.2. Theorem 2.1 holds for Problem II (with $W_0^{1,2}(\Omega)$ replaced by $W^{1,2}(\Omega)$).

For a further discussion of Problems I and II when Ω is a non-convex polygonal domain see section 5.

3. The Finite Element Spaces and Some of Their Properties.

Unless otherwise stated, in this section Ω will be a polygonal domain in $\mathbb{R}^2$. Our aim here is to define certain triangulations of Ω and discuss some properties of particular finite dimensional subspaces of $W^{1,2}(\Omega)$, the continuous piecewise linear finite element spaces, which will be used to approximate the solutions of the boundary value problems discussed in section 2. A brief discussion of some other finite element spaces and the case when Ω is not polygonal will be given at the end.

Let $0 < h \leq 1$ be a discretization parameter and for each such h let π_h denote a partition of Ω into disjoint triangles T_i, $i = 1, \ldots, N(h)$, such that i) The maximum length of a side of any triangle T_i is $\leq h$. ii) The common edges of any two adjoining triangles coincide. iii) $\overline{\Omega} = \bigcup_{i=1}^{N(h)} \overline{T}_i$. iv) The triangulation is quasi-uniform i.e. the area of any triangle $T_i \varepsilon \pi_h$ is bounded below by ch^2, where

C is independent of h.

Remark 3.1. Property iv) may be restated as follows: There exists a constant M and $m > 0$ and independent of h such that every triangle T_i is contained in a ball of radius Mh and contains a ball of radius mh. Roughly speaking the triangles in π_h are of the "same" size and do not degenerate in that the minimum angle of any triangle is bounded away from zero uniformly in h. Families of such triangulations are often called quasi-uniform.

We start with particular finite dimensional subspaces of $W^{1,2}(\Omega)$ and $W_0^{1,2}(\Omega)$ which will be used throughout these notes. Let $S^h(\Omega)$ denote the continuous functions on $\bar{\Omega}$ which are linear on each triangle and let $S_0^h(\Omega)$ be the subspace of S^h consisting of those functions which vanish on $\partial\Omega$. These spaces are called the finite element spaces of piecewise linear functions (relative to the triangulation π_h) and were introduced by Courant. Obviously a linear function on a triangle is uniquely determined by specifying its values at the vertices which we shall call nodes and which we assume ordered. Every $\phi \varepsilon S_h(\Omega)$ may be uniquely written in the form

$$(3.1) \qquad \phi(x) = \sum \alpha_i \phi_i(x),$$

where $\phi_i \varepsilon S_h(\Omega)$ is that basis function which is 1 at the ith "node" and zero at all other nodes (see fig (**1**)).

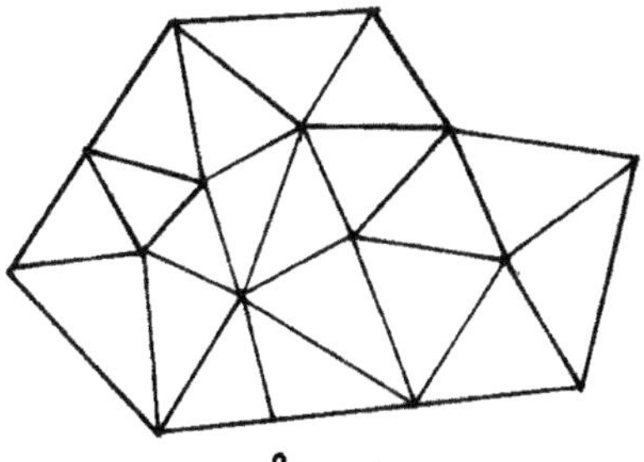

fig 1

For $S_0^h(\Omega)$ we only need those ϕ_i which are 1 at nodes which are interior points in Ω.

Since these functions are to be used to approximate the solutions of the boundary value problems stated in the previous section we will first discuss some of their approximation properties. In order to do so, let us first introduce the notion of the interpolant of a function.

Definition. Let $u \in C(\bar{\Omega})$, the interpolant $U_I \in S^h(\Omega)$ of u is defined to be the unique piece linear function which is equal to u at the nodal points of the triangulation.

Remark 3.2. U_I is not well defined for all $u \in W^{1,2}(\Omega)$ or $W_0^{1,2}(\Omega)$, since such u are not necessarily continuous on $\bar{\Omega}$. Sobolev's imbedding theorem gives some conditions under which it is well defined, for example if $u \in W^{2,1}(\Omega)$ or $u \in W^{1,p}(\Omega)$ for any $p > 2$.

Theorem 3.1. Let T be any triangle in π_h then

(i) If $u \in W^{2,2}(T)$

$$\|u-u_I\|_{L^2(T)} \leq Ch^2|u|_{2,2,T} , \tag{3.2}$$

$$\|\nabla(u-u_I)\|_{L^2(T)} \leq Ch|u|_{2,2,T} . \tag{3.3}$$

(ii) If $u \in C^2(\bar{T})$ then,

$$\|u-u_I\|_{C(\bar{T})} \leq Ch^j|u|_{C^j(\bar{T})} , \quad j = 0, 1, 2 \tag{3.4}$$

$$\|\nabla(u-u_I)\|_{C(\bar{T})} \leq Ch^{j-1}|u|_{C^j(\bar{T})} , \quad j = 1,2 \tag{3.5}$$

where C is independent of u, h and T.

Remark 3.3. If $u = P_1$ is a linear function on T then $u_I = P_1$, i.e. the interpolation operator is the identity on the subspace. The right hand sides of the above inequalities involve only semi-norms of u and in this case $|P_1|_{2,2,T} = |P_1|_{C^2(T)} = 0$.

An immediate consequence of Theorem 3.1 is the following:

Corollary 3.1. Let $\bar{D}_h$ be the union of the closures of any set of triangles in π_h.

(i) If $u \in W^{2,2}(D_h)$ (resp. $W_0^{1,2}(D_h) \cap W^{2,2}(D_h)$) then $U_I \in S^h(D_h)$ (resp. $S_0^h(D_h)$) satisfies

$$\|u-u_I\|_{L^2(D_h)} + h\,\|\nabla(u-u_I)\|_{L^2(D_h)} \leq Ch^2|u|_{2,2,D_h} . \tag{3.6}$$

(ii) If $u \in C^2(D_h)$ (resp $C^2(D_h) \cap C_0(D_h)$) then $u_I \in S^h(D_h)$ (resp $S_0^h(D_h)$) satisfies

(3.7) $$\|u-u_I\|_{C(\bar{D}_h)} + h\,\|\nabla(u-u_I)\|_{L^\infty(D_h)} \leq Ch^j\,|u|_{C^j(\bar{D}_h)},\ j = 1,2.$$

Where C is independent of u, h and D_h.

The proof of (3.7) is obvious from (3.4) and (3.5). The inequality (3.6) follows from (3.2) and (3.3) by squaring both sides, summing over all triangles T in D_h and then taking square roots.

Remark 3.4. There is a standard first step which is often used in proving various properties i.e. approximation properties, inverse properties (see Theorem 3.2) etc. of the subspaces $S^h(\Omega)$. This consists of mapping each of the triangles $T \in \pi_h$ onto a single reference T^* triangle, with vertices say (0,0), (1,0), (0,1), by means of orientation preserving affine mapping. Let x_1^*, x_2^* refer to the new coordinates on T^* where

$$x_1^* = \alpha_1 + \beta_1 x_1 + \gamma_1 x_2 \;.$$

$$x_2^* = \alpha_2 + \beta_2 x_2 + \gamma_2 x_2 \;,$$

for some constants α_1, β_1, γ_1, α_2, β_2, γ_2 which are completely determined by the assignment of the vertices of T to the vertices of T^*. It is important to note that a linear function on T is transformed into a linear function on T^*, in fact more generally a polynomial of degree $\leq r$ is transformed into polynomial of degree $\leq r$. The desired result is then hopefully proved on T^* (a unit size domain) and the result transformed back to T. The mapping from any such T to T^* may be thought of as a composition of two affine mappings. The first is a mapping onto a "reference" triangle T_h^* of size h, i.e. with vertices say (0,0), (h,0), (0,h), and then a simple scaling to T^*. Because of our assumption of quasi-uniformity the mapping from $T \to T_h^*$ and its inverse are "nice" with Jacobians uniformly bounded away from zero independent of T and h. Under these transformations for example

$$C_1\,\|u\|_{m,p;T} \leq \|u\|_{m,p,T_h^*} \leq C_2\,\|u\|_{m,p,T} \;,$$

with C_1 and C_2 independent of h and T and so in what follows we shall assume without loss of generality that $T = T_h^*$.

Proof of Theorem 3.1. In view of remark 3.4 we will prove the result with T replaced by T_h^* the "reference" triangle of size h. Transform T_h^* to T^* by the scaling

$$X_1^* = \frac{x_1}{h}, \quad x_2^* = \frac{x_2}{h}, \tag{3.8}$$

and let $u^*(x_1^*,x_2^*) = u(hx_1,hx_2)$ and $u_I^*(x_1,x_2) = u_I(hx_1^*,hx_2^*)$. Now

$$u_I^*(x_1^*,x_2^*) = u^*(0,0) + (u^*(1,0) - u^*(0,0)x_1 + (u^*(0,1) -u^*(0,0))x_2^*. \tag{3.9}$$

From this it easily follows that (using a crude bounds with the triangle inequality)

$$\|u_I^*\|_{L^2(T^*)} \le \|u_I^*\|_{C(T^*)} \le 5\|u^*\|_{C(T^*)}, \tag{3.10}$$

and

$$\|\nabla u_I^*\|_{L^2(T^*)} \le \| u_I^*\|_{C(T^*)} \le 5\|u^*\|_{C(T^*)}. \tag{3.11}$$

Let us now prove (3.2) and (3.3), the proof of (3.4) and (3.5) will be left to the reader. Since $u \in W^{2,2}(T^*)$ we have from (3.10) and Sobolev Inequality (1.5) that

$$\|u^*-u_I^*\|_{L^2(T^*)} \le \|u^*\|_{L^2(T^*)} + \|u_I^*\|_{L^2(T^*)} \le C\|u^*\|_{2,2,T^*},$$

where C is independent of u^* and h. Since the linear interpolant reproduces linear functions it follows that $(u^* - P^*)_1 = u_I^* - P^*$ where P^* is any linear function of the form

$$P^* = \alpha + \beta x_1^* + \gamma x_2^*.$$

Hence

(3.12) $$\|u^*-u_I^*\|_{L^2(T^*)} = \|(u^*-P^*)-(u^*-P^*)_I\|_{L^2(T^*)} \leq C\|u^*-P^*\|_{2,2,T^*} .$$

Similarly one derives.

(3.13) $$\|\nabla u^*-\nabla u_I^*\|_{L^2(T^*)} \leq C\|u^*-P^*\|_{2,2,T^*} .$$

As a next and crucial step we shall show that P^* may be chosen (depending on u^*) so that

(3.14) $$\|u^*-P^*\|_{2,2,T_h^*} \leq C|u^*|_{2,2,T^*} ,$$

where we emphasize that the term on the right only involves the semi-norm. Granting this last inequality, let us prove (3.2) by simply transforming back to T_h^*. We have using $x_1^* = \frac{x_1}{h}$, $x_2^* = \frac{x_2}{h}$

$$\int_{T^*} (u^*-u_I^*)^2 \, dx^* = h^{-2} \int_{T_h^*} (u-u_I)^2 \, dx$$

and

$$|u^*|_{2,2,T^*}^2 = \iint_{T^*} \left(\frac{\partial^2 u^*}{\partial x_1^* \partial x_1^*}\right)^2 + \left(\frac{\partial^2 u^*}{\partial x_1^* \partial x_2^*}\right)^2 + \left(\frac{\partial^2 u^*}{\partial x_2^* \partial x_2^*}\right)^2 dx^*$$

$$= \iint_{T_h^*} \left(\frac{\partial^2 u}{\partial x_1 \partial x_1}\right)^2 + \left(\frac{\partial^2 u}{\partial x_1 \partial x_2}\right)^2 + \left(\frac{\partial^2 u}{\partial x_2 \partial x_2}\right)^2 dx$$

$$= |u|_{2,2,T_h^*}^2 .$$

Combining (3.12) and (3.14) and then using the above equalities, (3.2) follows. A similar scaling using (3.13) and (3.14) proves (3.3). We now turn to the proof of (3.14).

Let $V = u^*-P^*$. We shall choose P^* so that

$$\int_{T^*} V\, dx^* = \int_{T^*} \frac{\partial V}{\partial x_1^*}\, dx^* = \int_{T^*} \frac{\partial V}{\partial x_2^*}\, dx^* = 0.$$

Now

$$\frac{\partial V}{\partial x_1^*} = \frac{\partial u^*}{\partial x_1^*} + \beta \quad , \quad \frac{\partial V}{\partial x_2^*} = \frac{\partial u^*}{\partial x_2^*} + \gamma ,$$

So the choices $\beta = - \int_{T^*} \frac{\partial u^*}{\partial x_2^*}\, dx^*$, $\gamma = - \int_{T^*} \frac{\partial u^*}{\partial x_2^*}\, dx^*$ and

$$\alpha = \int_{T^*} u^* - \beta \int_{T^*} x_1^*\, d_x^* - \gamma \int_{T^*} x_2^*\, dx^* \quad \text{accomplish this.}$$

Now apply Poincare's inequality (1.1) on T^* (in this case $d = 1$). Since $\int_{T^*} V\, dx^* = 0$.

$$(3.15) \qquad \|V\|_{L^2(T^*)} \leq \|\nabla V\|_{L^2(T^*)}$$

Since $\frac{\partial V}{\partial x_1^*}$ and $\frac{\partial V}{\partial x_2^*}$ also have mean value zero on T^* we may apply (1.1) to them separately and then adding

$$(3.16) \qquad \|\nabla V\|_{L^2(T^*)} \leq |V|_{2,2,T^*}$$

Since the semi-norm only contains second derivatives which annihilates any P^* we have from (3.15) and (3.16) that

$$\|u^*-P^*\|_{2,2,T^*} = \|V\|_{2,2,T^*} \leq C|V|_{2,2,T^*} = C|u^*|_{2,2,T^*}$$

which completes the proof.

Remark 3.3. The critical point in the proof of Theorem 3.1 occurs in (3.14). We used Poincare's inequality to prove this. Let us note that $\|u^*-P^*\|_{2,2,T^*}$ can be replaced by $\inf_{P^*}\|u-P^*\|_{2,2,T^*}$. This is just the norm on $W^{2,2}(T^*)/P^*$, i.e. the quotient space of $W^{2,2}(T^*)$ modulo the linear polynomials. The very useful Bramble-Hilbert Lemma states more generally that for m an integer $\inf_{P^*_{m-1}} \|u^*-P^*_{m-1}\|_{m,p,T^*}$, which is the norm on the quotient space of $W^{m,p}(T^*)$ modulo polynomials P^*_{m-1} of degree $\leq m-1$, is equivalent to the semi-norm $|u^*|_{m,p,T^*}$.

This can be used then to prove approximation results in other norms. For an excellent and detailed exposition of approximation properties of finite elements along these lines we refer the reader to Ciarlet [4].

We now turn to proving some other fundamental properties of the subspaces S^h. The first results belong to a class of useful inequalities usually referred to as inverse properties. We shall consider only those which will be needed in what follows.

Theorem 3.2. Let $\{\pi_h\}$ be a quasi-uniform family of triangulations and $S^h(\Omega)$ as above. Then there exists a constant C such that if $\bar{D}_h$ is the closure of the union of any set of triangles in π_h then

(3.17) $$\|\chi\|_{1,2,D_h} \leq Ch\|\chi\|_{L^2(D_h)},$$

(3.18) $$\|\chi\|_{C(\bar{D}_h)} \leq Ch^{-1}\|\chi\|_{L^2(D_h)},$$

where C is independent of h, χ and D_h.

Remark 3.4. These inequalities say that on the subspace S_h we may bounded "stronger norms" by "weaker norms". The analogous inequalities do not hold for example for if we replace $\chi \in S^h$ with $\chi \in W^{1,2}(D_h)$ or $C(D_h)$ respectively. Notice as $h \to 0$ the above estimates degenerate.

Proof of Theorem 3.2. We look at any triangle $T \in \pi_h$. In view of Remark 3.4, we shall again assume for simplicity that $T = T^*_h$. As in the proof of Theorem 3.1, T^*_h is transformed to T via $x^*_1 = \frac{x_1}{h}$, $x^*_2 = \frac{x_2}{h}$. Let χ^* be the transformed χ. Note now that the space of linear functions on T^*_h are mapped onto the space of linear functions on T^*, which is a fixed finite dimensional space (of dimension 3 in our special case) independent of h. Since all norms on a finite dimensional space are equivalent we have for any linear χ^* on T^*

(3.19) $$\|\chi^*\|_{1,2,T^*} \le C\|\chi^*\|_{L^2(T^*)}$$

(3.20) $$\|\chi^*\|_{C(\overline{T}^*)} \le C\|\chi^*\|_{L^2(T^*)},$$

and the results (3.17) follows easily by transforming (3.19) back to T_h^* and then T, squaring both sides and summing over all $T \in D_h$. The inequality (3.18) follows also by transforming back to T_h^*. We leave the details to the reader.

The next result we shall consider is known as a "superapproximation" property, which we shall prove in special form which will be convenient in the next section.

Theorem 3.3. Let $\overline{D}_h$ be the union of the closures of any set of triangles in π_h. Furthermore let $w(x,y) \in C^2(\overline{D}_h)$ and $\chi \in S^h(\overline{D}_h)$, then there exists a constant C such that

(3.21) $$\|\nabla(w^2\chi)-\nabla w\chi)_I\|_{L^2(D_h)} \le Ch(|w|_{C^1(\overline{D}_h)}\|w\nabla\chi\|_{L^2(\overline{D}_h)} + |w|_{C^2(\overline{D}_h)}\|\chi\|_{L^2})$$

where C is independent of h, D_h, χ and w.

Remark 3.5. If $w \equiv 1$ then the right hand side of (3.21) vanishes. The inequality (3.21) says that for products of functions of this type the interpolant yields an O(h) approximation but in contrast to (3.6) the terms on the right only involve the χ and its gradient in L^2. It is for this reason that this is usually referred to as the superapproximation property.

Proof: Let T be any triangle in π_h. Then using the approximation result (3.3).

(3.22) $$\|\nabla(w^2\chi)-\nabla(w^2\chi)_I\|_{L^2(T)} \le Ch^2|w\chi|_{2,2,T} \quad .$$

We now evaluate the right hand side in a little more detail. Using Leibnitz's rule and noticing that since χ is linear on T, so that

$$\frac{\partial^2\chi}{\partial x_1\partial x_2} = \frac{\partial^2\chi}{\partial x_1{}^2} = \frac{\partial^2\chi}{\partial x_2{}^2} = 0, \quad \text{then}$$

$$\frac{\partial^2 w^2\chi}{\partial x_i\partial x_j} = \frac{\partial w^2}{\partial x_i}\frac{\partial\chi}{\partial x_j} + \frac{\partial w^2}{\partial x_j}\frac{\partial\chi}{\partial x_i} + \frac{\partial^2 w^2}{\partial x_i\partial x_j}\chi, \quad i,j = 1,2.$$

$$= 2\frac{\partial w}{\partial x_i}w\frac{\partial\chi}{\partial x_j} + 2\frac{\partial w}{\partial x_j}w\frac{\partial\chi}{\partial x_i\partial x_j} + \frac{\partial^2 w}{\partial x_i\partial x_j}\chi \ .$$

From this it easily follows that

$$\left\|\frac{\partial^2(w\chi)}{\partial x_i\partial x_j}\right\|^2_{L^2(T)} \le C\left(|w|^2_{C^1(\bar{T})}\|w\nabla\chi\|^2_{L^2(T)} + |w|^2_{C^2(\bar{T})}\|\chi\|^2_{L^2(T)}\right)$$

From which the right hand side of (3.22) may be estimated. The result now follows by summing over the triangles in D_h.

We now consider one more technical result which will be used later on. Sobolev's Lemma 1.3 says that in two dimensions the maximum norm of a function defined on $\bar{\Omega}$ can be bounded by $C(p)\|u\|_{1,p,\Omega}$ for any $p > 2$, where $C(p)$depends on p. Emotionally speaking it can "almost" be bounded by the $W^{1,2}(\Omega)$ norm. However if we restrict u to the subspaces $S_0^h(\Omega)$ then the next result says that this can be done with a constant which blows up as $h \to 0$ in a very specific way.

<u>Theorem 3.4</u>. Let Ω be a polygonal domain and $S_0^h(\Omega)$ as above then there exists a constant C such that for any $\chi \in S_0^h(\Omega)$

(3.23) $$\|\chi\|_{C(\bar{\Omega})} \le C(\ln 1/h)^{1/2}\|\nabla\chi\|_{L^2(\Omega)} \quad .$$

where C is independent of h and χ.

We shall give a proof which depends on the fact that χ vanishes on $\partial\Omega$. Let $x_0 \in \bar{\Omega}$ be such that $\|\chi\|_{C(\bar{\Omega})} = |\chi(x_0)|$. Extend χ as zero outside of $\bar{\Omega}$ to a ball $B(x_0 d)$ of radius d such that $\bar{\Omega} \subset B(x_0,d)$. The function $G(x_0,x) = \frac{1}{2\pi} \ell n \left|\frac{x-x_0}{d}\right|$ is the Greens function on $B(x_0,d)$ for $-\Delta$ with singularity at x_0. Then

$$\chi(x_0) = \int_{B(x_0,d)} \nabla G(x_0 x)\ \nabla\chi(x)dx = \int_{B(x_0,h)} \nabla G(x_0 x)\nabla\chi(x)dx$$

$$+ \int_{B(x_0 d)/B(x_0,h)} \nabla G(x_0,x)\nabla\chi(x)dx = I_1 + I_2 .$$

Let $r = |x-x_0|$, then $\left|\frac{\partial G}{\partial x_i}\right| \leq \frac{C}{r}$, $i = 1,2$. Using polar co-ordinates

$$\|\nabla G\|_{L^2(B(x_0 h))} \leq C \int_0^h dr = ch$$

Hence using an inverse property on $S_0^h(\Omega)$.

$$|I_1| \leq \|\nabla G\|_{L^1(B(x_0 h))} \|\nabla\chi\|_{L^\infty(\Omega)} \leq Ch^{-1} \|\nabla G\|_{L^1(B(x_0 h))} \|\nabla\chi\|_{L^2(\Omega)}$$

$$\leq \|\nabla\chi\|_{L^2(\Omega)} .$$

Using the Cauchy-Schwarz inequality.

$$|I_2| \leq \|\nabla\chi\|_{L^2(\Omega)} \|\nabla G\|_{L^2(B(x_0,d)/B(x_0,h))}$$

$$\leq C\|\nabla\chi\|_{L^2(\Omega)} \left(\int_h^d \frac{1}{r}\, dr\right)^{1/2}$$

$$\leq C\left(\ell n\frac{1}{h}\right)^{1/2} \|\nabla\chi\|_{L^2(\Omega)}$$

Taken together these last two inequalities prove the result.

Let us first remark that this result still holds if $\chi \in S^h(\Omega)$ provided $\|\nabla\chi\|_{L^2(\Omega)}$ is replaced with $\|\chi\|_{1,2,\Omega}$ on the right hand side.

Let us end this section by considering some other possible choices of approximating functions and then the case where Ω is not a polygonal. We have discussed the situation where our approximating functions are continuous piecewise linear function on Ω a polygonal domain. For this space the interpolant is second order accurate (in h) in $L^2(\Omega)$ and $C(\bar{\Omega})$ provided $u \in W^{2,2}(\Omega)$ and $C^2(\bar{\Omega})$ respectively. This is the best that one can expect in general. Higher order accuracy can be obtained by using functions which are higher order polynomials on each triangle provided u is smoother than above. For example, let $S^{h,r}(\Omega)$ denote the space of functions which are continuous on $\bar{\Omega}$ and which are polynomials of degree $r-1$, $r \geq 2$, on each triangle $T \in \pi_h$. These are uniquely determined by specifying their values at nodal points. Rather than giving a general discussion of what these are let us illustrate them in two important cases.

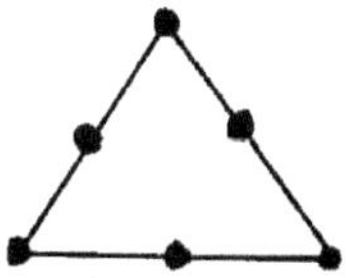

$S^{h,3}(\Omega)$

quadratic polynomial on T

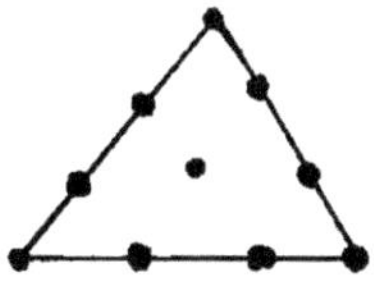

$S^{h,4}(\Omega)$

cubic polynomial on T

nodal points

For $S^{h,3}(\Omega)$ we take the vertices of T and the midpoints of each side as nodes and for $S^{h,4}(\Omega)$ these are the vertices of T, two equally spaced points on each side of T and one interior point. It is obvious then how to define the interpolant and following the proof of Theorem 3.1 one obtains that if $u \in W^{j,2}(\Omega)$ and $C^j(\bar{\Omega})$ respectively then

(3.24) $$\|u-u_I\|_{L^2(\Omega)} \leq Ch^j |u|_{j,2,\Omega} \qquad j = 2, \ldots, r,$$

(3.25) $$\|\nabla(u-u_I)\|_{L^2(\Omega)} \leq Ch^{j-1} |u|_{j,2,\Omega} \qquad j = 2, \ldots, r,$$

(3.26) $$\|u-u_I\|_{C(\bar{\Omega})} \leq Ch^j |u|_{C^j(\bar{\Omega})}, \quad j = 0, 1, \ldots, r$$

and

$$\|\nabla(u-u_I)\|_{C(\bar{\Omega})} \leq Ch^{j-1} |u|_{C^j(\Omega)} \qquad j = 1, 2, \ldots, r$$

Thus for example the error in the interpolant is of order h^r in $L^2(\Omega)$ and $C(\bar{\Omega})$ respectively provided $u \in W^{r,2}(\Omega)$ and $C^r(\bar{\Omega})$ respectively. There is a large difference between the accuracy obtained with a piecewise linear function, $O(h^2)$, and say a piecewise cubic $O(h^4)$. The "higher order" subspaces are desirable in approximating the solutions of boundary value problems which are sufficiently smooth even though they are more complicated to implement and require more computational effort.

Let us now consider a case where Ω is not a polygonal domain. For simplicity let Ω be a convex domain with say a C^∞ boundary. In this case we triangulate Ω as before. Let $\bar{\Omega}_h = \bigcup_{i=1} \bar{T}_i$ where we further require that the vertices which ly on $\partial\bar{\Omega}_h$ also ly on $\partial\Omega$ (see fig 3)

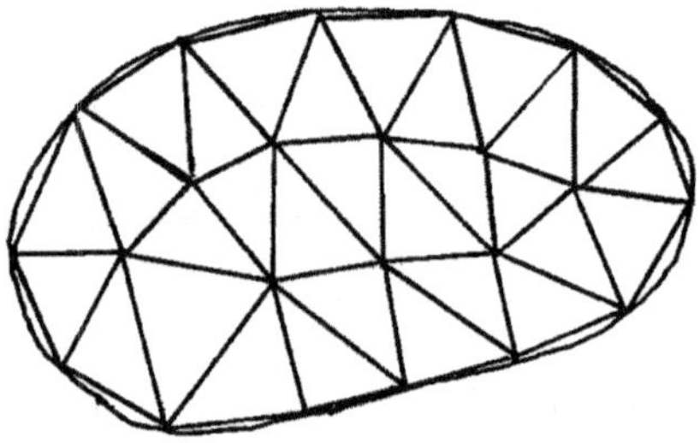

fig 3

Under these conditions $\Omega_h \subset \Omega$ and because we have triangulated with triangles with straight edges, roughly of size h, we are not able to fit the boundary exactly. However it is easily seen that

$$\text{dist}\,(\partial\Omega_h, \partial\Omega) \leq Ch^2 \tag{3.28}$$

Suppose we wished to approximate functions which vanish on $\partial\Omega$. We may take $S_0^h(\Omega)$ to be for example the continuous piecewise linear functions on $\bar{\Omega}_h$ which vanish on $\partial\Omega_h$ and simply extend them by zero outside of $\bar{\Omega}_h$. It is not hard to see that the estimates (3.24), (3.25) and (3.26) still hold (in the case $r = 2$). For higher order piecewise polynomial space there is difficulty in that (3.28) is not enough for (3.24), (3.25) and (3.26) to hold. This difficulty can be overcome by using "triangles" with curved sides which better fit the boundary and which are mapped onto a triangles with straight sides where the subspaces are taken to be the higher order polynomials. These so called isoparametric elements are discussed at length in [4] to which we again refer the reader.

4. The Galerkin method, The Finite Element method and Basic Error Estimates in $L^2(\Omega)$ and $W^{1,2}(\Omega)$.

Here we shall discuss the Galerkin method for approximating solutions of elliptic boundary value problems. For simplicity, this will be done here for Dirichlets' Problem (2.2). Roughly speaking one first chooses a finite dimensional subspace say S_N of $W_0^{1,2}(\Omega)$ whose functions can "closely" approximate the solution of the problem. The Galerkin approximation $u_N \in S_N$ is then characterized by the following:

Problem (IG) Find $u_N \in S_N$ satisfying

$$(4.1) \qquad (\nabla u_N, \nabla\phi) = (f,\phi) \quad \forall\ \phi \in S_N \ .$$

This amounts to discretizing the weak from the boundary value problem (2.2) using S_N instead of $W_0^{1,2}(\Omega)$. The "standard" finite element method is the Galerkin-method with special cases of S_N, usually spaces of piecewise polynomials. After discussing the Galerkin method error estimates will be derived when Ω is a polygonal domain and $S_N = S_0^h(\Omega)$.

Remark 4.1. A useful generalization of Galerkin method is the so called Petrov-Galerkin method. This is characterized by seeking an approximate solution u_N in one finite dimensional space say S_N (called the trial space) and requiring that (4.1) be satisfied for all ϕ in another finite dimensional space S_N^* (called the test space).

Let $\phi_1, \ldots, \phi_N$ be a basis for S_N. The problem (4.1) is equivalent to solving a system of N linear equation for the constants β_i, $i = 1, \ldots N$ where $u_N = \sum_{i=1}^{N} \beta_i \phi_i$. In fact substituting this into (4.1) and choosing $\phi = \phi_j$ $j = 1, \ldots N$ successively, we obtain

$$\sum_{i=1}^{N} \beta_i \ (\nabla\phi_i, \nabla\phi_j) = (f,\phi_j). \quad j = 1, \ldots N.$$

In matrix notation let A be the $N \times N$ matrix with entries $a_{ij} = (\nabla\phi_i, \nabla\phi_j)$, $\beta = (\beta_1, \ldots \beta_N)^T$ and $b = (b_1, \ldots b_N)^T$ with $b_j = (f,\phi_j)$ $j = 1, \ldots N$; then (4.1) becomes

$$(4.2) \qquad A\beta = b.$$

It is this equation which is solved in practice. The matrix A is a Gramm matrix relative to the inner product $(\nabla\cdot,\nabla\cdot)$ and is often called the stiffness matrix. In order to form this system of equations the inner products $(\nabla\phi_i,\nabla\phi_j)$ must be computed. It is important in practice that this be done without too much effort. If we take $S_N = S_0^h(\Omega)$ as discussed in Section 3 and $\phi_1, \ldots, \phi_N$ to be the nodal basis described there, the ϕ_i have small support and $(\nabla\phi_i,\nabla\phi_j) = 0$ corresponding to all nodes which are not connected by a single edge of a triangle. The resulting matrix is thus in general sparse.

Theorem 4.1. Let $f \in (W_0^{1,2}(\Omega))'$

(i) There exists a unique solution $u_N \in S_N$ of (4.1).

(ii) Furthermore if $u \in W_0^{1,2}(\Omega)$ satisfied (2.2) then u_N is the orthogonal projection of u onto S_N in the inner product $(\nabla\cdot,\nabla\cdot)$, i.e.

$$(4.3a) \qquad (\nabla(u-u_N),\nabla\phi) = 0 \quad \forall\ \phi \in S_N \quad .$$

Equivalently u_N is the closest element in S_N (in norm $\sqrt{(\nabla\cdot,\nabla\cdot)}$) to u i.e.

$$(4.3b) \qquad \|\nabla(u-u_N)\|_{L^2(\Omega)} = \inf_{\chi\in S_N} \|\nabla(u-\chi)\|_{L^2(\Omega)} \quad .$$

Remark 4.2. One basic property of the orthogonal projection is that it is bounded in norm i.e. in our case

$$(4.3c) \qquad \|\nabla u_N\|_{L^2(\Omega)} \leq \|\nabla u\|_{L^2(\Omega)} \quad .$$

This follows immediately from (4.3a) by choosing $\phi = u_N$, then

$$(\nabla u_N,\nabla u_N) = (\nabla u,\nabla u_N)$$

or

$$\|\nabla u_N\|^2_{L^2(\Omega)} \leq \|\nabla u\|_{L^2(\Omega)} \|\nabla u_N\|_{L^2(\Omega)} \quad ,$$

which implies (4.3c). Note also that if $u \in S_N$ then $u_N = u$.

Proof: Since by Corollary 3.1, $(\nabla\cdot,\nabla\cdot)$ is an inner-product on $W_0^{1,2}(\Omega)$ it is by restriction also one on S_N which is a Hilbert space. Since (f,ϕ) is a bounded linear functional on $W_0^{1,2}(\Omega)$, the Riesz-Representation Theorem implies the existence of a unique solution of (4.1). Note that since the ϕ_i are linearly independent, this implies that the equivalent matrix problem (4.2) has a unique solution.

Taking $\phi \in S_N$ in (2.2) and subtracting (4.2) from (2.1) we have

$$(4.4) \qquad (\nabla(u-u_n),\nabla\phi) = 0 \quad \forall \quad \phi \in S_N ,$$

thus u_N is the orthogonal projection of u as claimed. It is for this reason that the Galerkin method is often called a projection method. The inequality (4.3) is just the well known property of the orthogonal projection. To see this we use (4.4) to find for any $\chi \in S_N$

$$\|\nabla(u-u_N)\|^2_{L^2(\Omega)} = (\nabla(u-u_N),\nabla(u-u_N)) = (\nabla(u-u_N)\nabla u)$$

$$= (\nabla(u-u_N),\ \nabla u-\chi) \leq \|\nabla(u-u_N)\|_{L^2(\Omega)} \|\nabla(u-\chi)\|_{L^2(\Omega)}$$

where we have also used the Cauchy-Schwartz inequality. The inequality (4.3b) follows, which completes the proof.

The estimate (4.3b) says that we may obtain an upper bound for the error in $W_0^{1,2}(\Omega)$ in the Galerkin method by estimating the error in $W_0^{1,2}(\Omega)$ between u and any χ in the subspace S_N. We now take $S_N = S_o^h(\Omega)$, the finite element space of piecewise linear function on a quasi-uniform triangulation of Ω which from now on we shall assume to be a polygonal domain and prove some basic estimate in $W^{1,2}(\Omega)$ (energy estimates) for the finite element method.

Theorem 4.2. Let Ω be a polygonal domain and $u \in W_0^{1,2}(\Omega)$ satisfy (2.2). Let $u_h \in S_0^h(\Omega)$ be the finite element solution of (4.1) i.e.

$$(4.5) \qquad (\nabla u,\nabla\phi) = (f,\phi) \quad \forall\ \phi \in W_0^{1,2}(\Omega)$$

and

$$(4.6) \qquad (\nabla u_h,\nabla\phi) = (f,\phi) \quad \forall\ \phi \in S_0^h(\Omega).$$

(i) If $u \in W^{2,2}(\Omega)$, then

$$\|\nabla(u-u_h)\|_{L^2(\Omega)} \le Ch|u|_{W^{2,2}(\Omega)} \le Ch\|f\|_{L^2(\Omega)} . \tag{4.7}$$

(ii) If Ω is convex and $f \in L^2(\Omega)$ then $u \in W^{2,2}(\Omega) \cap W_0^{1,2}(\Omega)$ and (4.7) holds.

(iii) If Ω is non-donvex with maximal interior angle $\pi < \alpha_m \le 2\pi$, and $f \in L^p(\Omega)$, with $p = \frac{2\alpha_m}{2\alpha_m - \pi}$, then $u \in W_0^{2,p-\varepsilon}(\Omega)$ for any $\varepsilon > 0$ and

$$\begin{aligned} \|\nabla(u-u_n)\|_{L^2(\Omega)} &\le C_\varepsilon h^{(\pi/\alpha_m)-\varepsilon} |u|_{2,p-\varepsilon,\Omega} \\ &\le C_\varepsilon h^{(\pi/\alpha_m)-\varepsilon} \|f\|_{L^p} \end{aligned} \tag{4.8}$$

for any $\varepsilon > 0$.

In the above the constants C are independent of h, u and u_h.

<u>Proof</u>: If we assume that $u \in W^{2,2}(\Omega)$ then it immediately follows from (3.6), with the choice of $\chi = u_I$ and $D_h = \Omega$, that

$$\|\nabla(u-\chi)\|_{L^2(\Omega)} \le Ch|u|_{2,2,\Omega} ,$$

(4.7) now follows from (4.3b).

In fact if $f \in L^2(\Omega)$ and Ω is a convex polygon then from Theorem 2.1, $u \in W^{2,2}(\Omega)$ and the above assumption is valid. In the non-convex case we cannot guarantee this (even if $f \in C^\infty(\bar{\Omega})$). In general for a non-convex domain $u \in W^{2,p-\varepsilon}(\Omega)$ for any $\varepsilon > 0$, where, if $\pi < \alpha_m \le 2\pi$ is the maximal angle, $p = \frac{2\alpha_m}{2\alpha_m - \pi} < 2$. One can prove (using the Bramble-Hilbert Lemma (cf. Remark 3.2) that if $u \in W^{2,q}(\Omega)$ for $1 \le q \le 2$ then

$$\|u-u_I\|_{L^2(\Omega)} \le Ch^{3-2/q}|u|_{2,q,\Omega} \tag{4.9}$$

$$\|\nabla u-\nabla u_I\|_{L^2(\Omega)} \le Ch^{2-2q}|u|_{2,q,\Omega} \tag{4.10}$$

The estimates (4.10) and (2.4) in conjunction with (4.3) yield (4.8), which completes the proof.

The energy estimates given above for the finite element method are optimal with respect to the power of h in the sense that they are the best that we can expect in general from this subspace from approximation theory even if the solution were explicitly known apriore. We now ask how well does the approximate solution approximate u in other norms, for example in $L^2(\Omega)$ (maximum norm estimates are discussed in section 6)? For a convex domain and $f \in L^2(\Omega)$, $u \in W^{2,2}(\Omega)$ and (3.6) says that for the interpolant, $\|u-u_I\|_{L^2(\Omega)} \leq Ch^2|u|_{2,2,\Omega}$.

We shall show that this estimate also holds with the finite element solution u_h replacing u_I. As discussed previously if Ω is non-convex, with $\pi < \alpha_m \leq 2\pi$, and $f \in L^p(\Omega)$ then since $u \in W^{2,p-\varepsilon}(\Omega)$, $\forall\ \varepsilon > 0$ and p given in (4.8), (4.9) yields $\|u-u_I\|_{L^2(\Omega)} \leq C_\varepsilon h^{1+\frac{\pi}{\alpha_M}-\varepsilon}|u|_{2,p-\varepsilon,\Omega}$.

Unfortunately this estimate does not seem to hold with u_I replace by the finite element solution u_h. The best that has been proven so far (even if $f \in C^\infty(\Omega)$) is

$$\|u-u_h\|_{L^2(\Omega)} \leq C_\varepsilon h^{2(\frac{\pi}{\alpha_m}-\varepsilon)}|u|_{2,p-\varepsilon,\Omega} .$$

Notice that since $\pi/\alpha_m < 1$ $\frac{2\pi}{\alpha_m} < 1 + \frac{\pi}{\alpha_M}$. Numerical experiments hint that this estimate is sharp with respect to the power of h. Both of these estimates will be proved using a fundamental duality argument due to Aubin and Nitsche.

Theorem 4.3. Under the conditions of Theorem 4.2

(i) If Ω is convex and $f \in L^2(\Omega)$, then

$$(4.11)\qquad \|u-u_h\|_{L^2(\Omega)} \leq Ch^2|u|_{2,2,\Omega} \leq Ch^2\|f\|_{L^2(\Omega)} .$$

(ii) If Ω is non-convex and $f \in L^p$ with α_m and p as in (4.6) then

$$(4.12)\qquad \|u-u_n\|_{L^2(\Omega)} \leq C_\varepsilon h^{2(\frac{\pi}{\alpha_m}-\varepsilon)}|u|_{2,p-\varepsilon,\Omega} \leq Ch^{2(\frac{\pi}{\alpha_m}-\varepsilon)}\|f\|_{L^p(\Omega)}$$

Proof: Now

(4.8) $$\|u-u_h\|^2_{L^2} = (u-u_h, u-u_h)$$

We know that $u - u_h$ satisfies (4.4). The idea now is to connect the L^2 with the $W_0^{1,2}(\Omega)$ inner product. This is essentially done by introducing a function v which is the "weak" solution of $-\nabla v = u-u_h$ in Ω, $v = 0$ on $\partial\Omega$ and then integrating (4.8) by parts. Equivalently let $v \in W_0^{1,2}(\Omega)$ satisfy

(4.9) $$(\nabla v, \nabla\phi) = (u-u_h, \phi) \quad \forall\ \phi \in W^{1,2}(\Omega).$$

Now $u-u_h \in L^2(\Omega)$ and hence if Ω is convex $u \in W^{2,2}(\Omega) \cap W_0^{1,2}(\Omega)$ and

(4.10) $$\|v\|_{2,2,\Omega} \leq C\ \|u-u_h\|_{L^2(\Omega)}$$

If Ω is non convex then we are only able to say that, with p as above,

(4.11) $$\|v\|_{2,p-\varepsilon,\Omega} \leq C\ \|u-u_h\|_{L^p(\Omega)} \leq C\ \|u-u_h\|_{L^2(\Omega)}$$

for any $\varepsilon > 0$.

Using (4.9) with $\phi = u-u_h$ and the basic orthoginality property (4.4) of $u-u_h$

(4.12) $$(u-u_h, u-u_h) = (\nabla(u-u_h), \nabla v) = (\nabla(u-u_h), \nabla(v-\chi)).$$

$$\leq \|\nabla(u-u_h)\|_{L^2(\Omega)}\ \|\nabla(v-\chi)\|_{L^2(\Omega)}.$$

for any $\chi \in S_0^h(\Omega)$. Choosing $\chi = V_h$ the finite element approximation of V (or $\chi = V_I$ the interpolant of V), we obtain from (4.7), (4.10) and (4.12) when Ω is convex

$$\|u-u_h\|^2_{L^2(\Omega)} \leq Ch^{2(\frac{\alpha}{\pi_m} - \varepsilon)}\ |u|_{2,p-\varepsilon,\Omega}\ |V|_{2,p-\varepsilon,\Omega}$$

$$\leq Ch^{2(\frac{\pi}{\alpha_m} - \varepsilon)}\ |u|_{2,p-\varepsilon,\Omega}\ \|u-u_h\|_{L^2(\Omega)}$$

which completes the proof of the Theorem.

Remark 4.3. Instead of starting the proof of Theorem 4.2 with (4.8) we could have started with characterizing the $L^2(\Omega)$ norm of $u-u_h$ by

$$\|u-u_h\|_{L^2(\Omega)} = \sup_{\psi\neq 0} \frac{(u-u_h,\psi)}{\|\psi\|_{L^2(\Omega)}}$$

For each such ψ let V satisfy (4.9) with $(u-u_h)$ replacing $u-u_h$. The proof proceeds then as before and instead of (4.12) we have

$$\frac{(u-u_h,\psi)}{\|\psi\|_{L^2(\Omega)}} \le \|\nabla(u-u_h)\|_{L^2(\Omega)} \frac{\|\nabla(V-\chi)\|_{L^2(\Omega)}}{\|\psi\|_{L^2(\Omega)}}$$

The gain in order of accuracy in L^2 as opposed to $W_0^{1,2}(\Omega)$ again depends on the error in the L^2 norm of the gradient for the solution of the "adjoint problem" (see below). This gain is essentially limited, independent of the piecewise polynomial space used, by the fact that (for example for a convex polygonal or smooth domain

$$\|V\|_{2,2,\Omega} \le C\|\psi\|_{L^2(\Omega)}$$

So that we are limited to estimating $\|\nabla(V-\chi)\|_{L^2(\Omega)}$ by a term on the right which involves at most $\|V\|_{2,2,\Omega}$.

Remark 4.4. If instead of the Laplacian we were to consider a more general second order elliptic operator L, which is coercive on $W_0^{1,2}(\Omega)$ (see Showalter [ ]) then a similar argument holds. In this case the auxiliary problem introduced in the proof of Theorem 4.3 is replaced with the weak form of the adjoint problem $L^*V = u-u_h$ in Ω, $V = 0$ on $\partial\Omega$.

5. Local Error Estimates.

As discussed in the previous section the finite element solution of (4.6) is the orthogonal projection of the solution $u \in W_0^{1,2}(\Omega)$ of (4.5) into $S_0^h(\Omega)$. In general therefore it depends on the properties of u in all of Ω. Along these lines Theorem 4.2 gives estimates for the error which depend on the smoothness of u in all of Ω. In the case that Ω is non-convex and we are again using the continuous piecewise linear subspaces $S_0^h(\Omega)$, we obtained for example

$$\|\nabla(u-u_h)\|_{L^2(\Omega)} \leq C_\varepsilon h^{(\pi/\alpha_m)-\varepsilon} |u|_{2,p-\varepsilon,\Omega}$$

where $p = \frac{2\alpha_m}{2\alpha_m - \pi}$ and $\pi < \alpha_m \leq 2\pi$. Again note that $\pi/\alpha < 1$ and in general we cannot improve this estimate even if $f \in L^2(\Omega)$ or $C^\infty(\bar{\Omega})$. Let us simplify matters a little by assuming that all the corners are convex except one, i.e. $0 < \alpha_1 \leq \alpha_2 \leq \dots \leq \alpha_{m-1} < \pi$ and $\pi < \alpha_m \leq 2\pi$. Then $u \in W^{2,p-\varepsilon}(\Omega)$ as before however it turns out that $u \in W^{2,2}(\tilde{\Omega})$ on any subdomain of $\tilde{\Omega}$ which excludes a neighborhood of the corner with interior angle α_m. Here we are assuming that $f \in L^2(\Omega)$ (see Grisvard [6]). It is natural to ask whether the accuracy of the finite element solution is better in the region away from the corner where u is smoother? In this particular case the answer is yes; in fact the error will be of optimal order h except in the case when $\alpha_m = 2\pi$ when it is of order $h^{1-\varepsilon}$, any $\varepsilon > 0$. Another interesting situation occurs even when the boundary of Ω is smooth but f and hence u is not smooth in certain parts of Ω and smooth in others and again the same question may be asked. In this section we shall derive local error estimates for the finite element method i.e. estimates on subdomains of Ω and see to what extend they are influenced by the solution u and the nature of the domain away from the subdomain. These estimates have been useful in a number of applications. They will also be useful in our derivation of maximum norm estimates given in the following section. For simplicity in presentation we shall give a slightly modified treatment of a special case of a result first proved in Nitsche-Schatz [10].

We start with some notation. For $x \in \bar{\Omega}$, $B(x,d)$ will denote the open ball of radius d centered at x and $B'(x,d) = B(x,d) \cap \Omega$. The basic error estimate is as follows:

<u>Theorem 5.1</u>. Let u and u_h satisfy (4.5) and (4.6) respectively. There exist positive constants C and C^* such that if h is sufficiently small and $d \geq C^*h$ then for any $\chi \in S_0^h$

$$(5.1) \quad \|\nabla(u-u_h)\|_{L^2(B'(x,2d))} \leq C \; (\|\nabla(u-\chi)\|_{L^2(B'(x,2d)} + d^{-1}$$

$$\|u-\chi\|_{L^2(B'(x,2d))} + d^{-1} \|u-u_h\|_{L^2(B'(x,2d))}$$

Here C and C^* are independent of u, h, u_h, x, χ and d.

The proof of this result is lengthy. Let us first discuss its meaning, state a generalization, and then give an application to a specific problem. The estimate (5.1) may be interpreted as follows: The error in the norm $W^{1,2}(B'(x,d))$ can be bounded by three terms. The first two are just the best approximation error that the subspace S_0^h can provide on the larger subdomain $B'(x,2d)$ (which for example can be bounded by the local interpolation error in $W^{1,2}(B'(x,2d))$. Very roughly speaking the first term is like the error in an $W^{1,2}$ projection of u just on $B'(x,2d)$. The third term is the error measured in the "weaker" norm on $L^2(B'(x,2d))$, which is multiplied by the inverse of the distance between the two domains. The effects on the error due to the nature of the boundary or solution outside of $B'(x,d)$ are still present in this latter term which must be estimated separately for each particular problem. One way to do this is first use the inequality

$$\|u-u_h\|_{L^2(B'(x,2d))} \leq \|u-u_h\|_{L^2(\Omega)} .$$

and then estimate this latter term by the methods of the previous section. The significance of this weaker norm is that, under some important circumstances one prove higher rates of convergence in $L^2(\Omega)$ with relatively less requirements on the smoothness of u then one would need, for example in order to obtain the same rate of convergence in the $W^{1,2}$ norm. Let us remark that the L^2 norm on the right hand side of (5.1) may be replaced but much "weaker" norms see e.g. [10].

Let us also remark that this result may be easily generalized to include more general subdomains of Ω. We shall do this now for two classes of subdomains the first of which will be useful in proving maximum norm estimates and the second so that we can give an easy application to the problem with a non-convex polygonal domain mentioned above.

Consider the concentric annuli

$$A_1 = \{x:\tfrac{d}{2} < |x-x_0| < \tfrac{3d}{2}\}, \quad A_2 = \{x:\tfrac{d}{4} \leq |x-x_0| \leq 2d\}$$

$$\text{and } A_1' = A_1 \cap \Omega, \quad A_2' = A_2 \cap \Omega.$$

<u>Corrollary 5.1.</u> The results of Theorem 5.1 hold with A_1' and A_2' replacing $B'(x,d)$ and $B'(x,2d)$ respectively where C is independent of u, h, u_h, x_0, χ and d.

Proof: Cover A_1 with a fixed number (independent of d) of balls $B(x,\frac{d}{8})$ of radius say d/8. Then also consider the balls $B(x_i,\frac{d}{4})$. Apply Theorem 5.1 to each pair $B'(x_i, \frac{d}{8})$ and $B'(x_i, \frac{d}{4})$ the desired result is easily obtained but appropriately summing.

Now let $\Omega_i \subset \Omega_2$ be any fixed subdomains of Ω. Let $\Gamma_1 = \partial\Omega_1/(\partial\Omega, \cap \partial\Omega)$, $\Gamma_2 = \partial\Omega_2/(\partial\Omega_2 \cap \partial\Omega)$ i.e. Γ_1 and Γ_2 are the parts of $\partial\Omega_1$ and $\partial\Omega_2$ respectively which are not common to $\partial\Omega$. Suppose the $\text{dist}(\Gamma_1,\Gamma_2) \geq d_0$ for some fixed d_0. Then

Corollary 5.2. With Ω_1 and Ω_2 as above Theorem 5.1 holds with Ω_1 and Ω_2 replacing $B'(x,d)$ and $B'(x,2d)$ respectively however C in general depends on Ω_0 and Ω_1.

We leave the proof to the reader.

Before proving Theorem 1 let us apply corollary 5.2 to the following concrete problem.

Example. Let Ω be the Slit domain given in figure ( ), and let u and u_h satisfy (4.5) and (4.1) respectively.

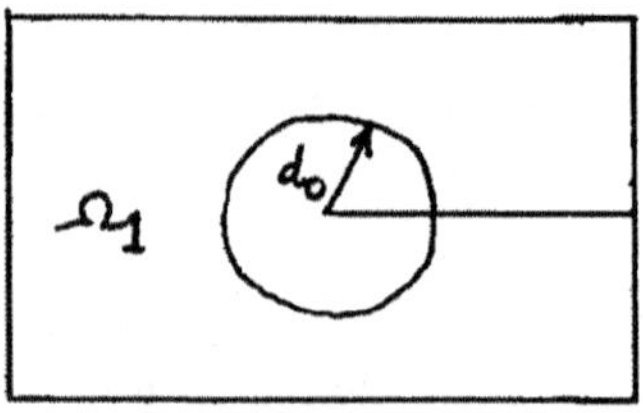

In this case the results of Theorems 4.2 and 4.3 yield the global estimates.

$$\|\nabla(u-u_h)\|_{L^2(\Omega)} \leq C_\varepsilon\, h^{1/2-\varepsilon}\, |u|_{2,\frac{4}{3}-\varepsilon,\Omega} \tag{5.2}$$

$$\|(u-u_h)\|_{L^2(\Omega)} \leq C_\varepsilon\, h^{1-2\varepsilon}\, |u|_{2,\frac{4}{3}-\varepsilon,\Omega} \; . \tag{5.3}$$

for any $\varepsilon > 0$

Let us apply Corollary 5.2 where we take $\Omega_1 = \Omega / B(0,d_0)$, $\Omega_2 = \Omega / B(0, do/2)$ for any fixed $d_0 > 0$. Now $u \in W^{2,2}(\Omega_3)$, where $\Omega_3 = \Omega / B(0,do/4)$ (see Grisvard [6]).

For h sufficiently small there exist a mesh domain $D_h, \Omega_2 \subset \bar{D}_h \subset \Omega_3$. Let $\chi = U_I$ and apply (3.6) to obtain

$$\|\nabla(u-\chi)\|_{L^2(\Omega_2)} + d_0^{-1}\|u-\chi\|_{L^2(\Omega_2)} \leq ch\,|u|_{2,2,\Omega_3}$$

Combining this estimate with (5.3) and corollary 5.2 yields the result

$$(5.4)\qquad \|\nabla(u-u_n)\|_{L^2(\Omega_1)} \leq C_\varepsilon h^{1-2\varepsilon}(|u|_{2,2,\Omega_3} + |u|_{2,\frac{4}{3}-\varepsilon,\Omega})$$

for any $\varepsilon > 0$. Comparing this with (5.2) this shows convergence for the gradients in $L^2(\Omega_1)$ in the finite element method is better away from the corners than on the whole domain. In the exercise 5.1 the reader is invited to generalize this example to other situations with non-convex corners.

<u>Proof of Theorem (5.1)</u>. The proof can be subdivided into two steps. First we will locally project u into the subspace. The first two term on the right of (5.1) are essentially a bound for the error in this local projection. We then obtain an estimate for the difference between this local projection and u_h, which has certain nice properties.

For technical reasons we shall need an increasing set of 5 domains Ω_j defined by $\Omega_j = B'(x,(1 + j/4)d)$ $j = 0, \ldots. 4$. Note that $\Omega_o = B'(x,d)$ and $\Omega = B'(x,2d)$. We localize u by employing a smooth "cut-off" function w satisfying $w \in C_0^\infty(\Omega_3)$, $w \geq 0$, $w \equiv 1$ on Ω_2 and $|D^\alpha w| \leq Cd^{-|\alpha|}$. Such a function is easily constructed. Consider the function wu which vanishes outside of Ω and let $P(wu)\in S_0^h(\Omega_4)$ be the projection wu in $W_0^{1,2}(\Omega_4)$ defined by

$$(5.5)\qquad (\nabla wu - \nabla P(wu), \nabla\phi) = 0 \quad \forall\ \phi \in S_0^h(\Omega_4).$$

Since $w \equiv 1$ on Ω_2,

$$(5.6)\qquad \|\nabla u-\nabla u_h\|_{L^2(\Omega_0)} \equiv \|\nabla(wu) - \nabla u_h\|_{L^2(\Omega_0)}$$

$$\leq \|\nabla wu - \nabla P(wu)\|_{L^2(\Omega_0)} + \|\nabla P(wu) - \nabla u_h\|_{L^2(\Omega_0)}$$

$$= I_1 + I_2.$$

To estimate I_1 we use (4.3) with the choice $\chi = 0$, and the properties of w to obtain

$$(5.7)\quad I_1 \le \|\nabla(wu)-\nabla P(wu)\|_{L^2(\Omega_4)} \le C\|\nabla(wu)\|_{L^2(\Omega_3)}$$

$$\le C(\|\nabla u\|_{L^2(\Omega_3)} + d^{-1}\|u\|_{L^2(\Omega_3)}).$$

Set $V_h = P(wu) - \nabla u_h$. In order to estimate I_2 first notice that since u_h satisifes

$$(5.8)\qquad (\nabla u-\nabla u_h,\nabla\phi) = 0 \qquad \forall\ \phi \in S_0^h(\Omega)$$

and $wu = u$ on Ω_2 then subtracting (5.8) from (5.5) yields

$$(\nabla V_h,\nabla\phi) = 0 \qquad \forall\ \phi \in S_0^h(\Omega)\ .$$

Such a $V_h \varepsilon S_0^h(\Omega)$ is called a "discrete harmonic function" on Ω. As we shall see it has certain properties similar to those of a harmonic function. In particular we shall show in our situation that

Lemma 5.1. Let V_h be as above, then there exists a constant C such that

$$(5.9)\qquad \|\nabla V_h\|_{L^2(\Omega_0)} \le C\, d^{-1}\|V_h\|_{L^2(\Omega_2)}\ ,$$

where C is independent of h, V_h, x and d.

Assuming that we have proved this result for a moment; let us complete the proof of (5.1). Using (5.9) and the triangle inequality

$$(5.10)\qquad I_2 \le C\, d^{-1}\|P(wu) - u_h\|_{L^2(\Omega_2)}$$

$$\le C\, d^{-1}(\|P(wu)\|_{L^2(\Omega_2)} + \|u-u_n\|_{L^2(\Omega_2)} + \|u\|_{L^2(\Omega_2)})$$

Since $P(wu) \varepsilon S_0^h(\Omega_4) \subset W_0^{1,2}(\Omega_4)$ where diam $(\Omega_4) \le 4d$, then using Poincare's inequality (1.4) and the properties of $P(wu)$.

$$(5.11)\qquad d^{-1}\|P(wu)\|_{L^2(\Omega_4)} \le C\|\nabla\ P(wu)\|_{L^2(\Omega_4)} \qquad C\|\nabla(wu)\|_{L^2(\Omega_3)}$$

$$\le C(\|\nabla u\|_{L^2(\Omega_3)} + d^{-1}\|u\|_{L^2(\Omega_3)})$$

Combining (5.11), (5.10), (5.7) and (5.6) we arrive at

$$(5.12)\qquad \|\nabla(u-u_h)\|_{L^2(\Omega_0)} \le C(\|\nabla u\|_{L^2(\Omega_4)} + d^{-1}\|u\|_{L^2(\Omega)} + d^{-1}\|u-u_h\|_{L^2(\Omega)}).$$

Now notice that for any $\chi \in S_0^h(\Omega)$

$$(\nabla u-\nabla u_h, \nabla\phi) = (\nabla(u-x) - \nabla(u_h-\chi), \nabla\phi) = 0\ ,$$

i.e. $u_h - \chi$ is the projection of $u - \chi$ and we apply the estimate (5.12) to this pair obtaining

$$\|\nabla u-\nabla u_h\|_{L^2(\Omega_0)} \le C(\|\ (u-\chi)\|_{L^2(\Omega_4)} + d^{-1}\|u-\chi\|_{L^2(\Omega_4)} + d^{-1}\|u-u_h\|_{L^2(\Omega_4)})$$

which is (5.1). Thus only the proof of (5.9) remains.

Proof of Lemma 5.1. Let ψ be a smooth "cut-off" function satisfying $\psi \in C_0^\infty(\Omega_1)$, $\psi \ge 0$, $\psi \equiv 1$ on Ω_0 and $|D^\alpha\psi| \le Cd^{-|\alpha|}$ for $|\alpha| \le 2$. Then

$$(5.13)\qquad \|\nabla V_h\|^2_{L^2(\Omega_0)} \le \|\psi\nabla V_h\|^2_{L^2(\Omega_1)} = \int_\Omega \psi^2\ \nabla V_h \cdot \nabla V_h\ dx$$

$$= \int_\Omega \nabla V_h \cdot \nabla(\psi^2 V_h)dx - \int_\Omega \nabla V_h \cdot 2\psi\ V_h\nabla\psi dx$$

$$= J_1 + J_2$$

Since ψ vanishes outside of Ω_1,

$$(5.14)\quad |J_2| = \left|\int_\Omega \psi\nabla V_h \cdot 2V_h\,\nabla\psi\,dx\right| \le \|\psi\nabla V_h\|_{L^2(\Omega_1)} \cdot Cd^{-1}\|V_h\|_{L^2(\Omega_1)}$$

$$\le \tfrac{1}{4}\|\psi\nabla V_h\|^2_{L^2(\Omega_1)} + Cd^{-2}\|V_h\|^2_{L^2(\Omega_1)},$$

where we have first used the Cauchy-Schwarz inequality and then the arithmatic-geometric mean inequality.

In order to estimate J_2 we note that, by taking C^* and hence d sufficiently large, there exists a mesh domain D_h (i.e. $\bar{D}_h = \cup\bar{T}_i$, $T_i \in \pi_h$) such that $\Omega_1 \subset D_h \subset \Omega_2$. The interpolant $(\psi^2 V_n)_I \in S_0^h(D_h) \subset S_0^h(\Omega_2)$ and since V_n satisfies (5.9),

$$|J_1| = \left|\int_\Omega \nabla V_h \cdot \nabla(\psi^2 V_h - (\psi^2 V_h)_I)\,dx\right| \le \|\nabla V_h\|_{L^2(D_h)}\,\|\nabla(\psi^2 V_h - (\psi^2 V_h)_I)\|_{L^2(D_h)}$$

Applying the "superapproximation" property (3.21) to the second term on the right it follows that

$$(5.15)\quad |J_1| \le C\|\nabla V_h\|_{L^2(D_h)}\, h(d^{-1}\|\psi\nabla V_h\|_{L^2(D_h)} + d^{-2}\|V_h\|_{L^2(D_h)}).$$

Using the arithmatic-geometric mean inequality and then the inverse property (3.14)

$$\|\psi\nabla V_h\|_{L^2(D_n)}\, Chd^{-1}\|\nabla V_h\|_{L^2(D_n)} \le \tfrac{1}{4}\|\psi\nabla V_h\|^2_{L^2(D_n)} + Ch^2 d^{-2}\|\nabla V_h\|^2_{L^2(D_n)}$$

$$\le \tfrac{1}{4}\|\psi\nabla V_h\|^2_{L^2(D_n)} + Cd^{-2}\|V_h\|^2_{L^2(D_n)}.$$

Again by the inverse property (3.19)

$$Chd^{-2}\|\nabla V_h\|_{L^2(D_n)}\,\|V_h\|_{L^2(D_h)} \le Cd^{-2}\|V_h\|^2_{L^2(D_n)}$$

Using these last two inequalities in (5.15) and combining the result with (5.14) and (5.13) we obtain

$$\|\psi\nabla v_h\|^2_{L^2(D_h)} \le \frac{1}{2}\|\psi\nabla v_h\|^2_{L^2(D_h)} + C\, d^{-2}\|v_h\|^2_{L^2(D_h)} .$$

Therefore

$$\|\nabla v_h\|_{L^2(\Omega_0)} \le \|\psi\nabla v_h\|^2_{L^2(D_h)} \le Cd^{-2}\|v_h\|^2_{L^2(D_h)} \le Cd^{-2}\|v_h\|^2_{L^2(\Omega_2)}$$

which completes the proof.

6. Error estimates in the maximum norm.

There are several methods which have been developed for proving sharp maximum norm error estimates for the finite element method for elliptic boundary value problems when the domain Ω has been triangulated with a quasi-uniform mesh. We shall first state the result which will be proved and then briefly discuss the ideas behind two of these methods. A modification of one of them will then be used to prove the result. Again for simplicity we shall limit ourselves here to studying Dirichlet's Problem I on a convex polygonal domain. Generalizations will be discussed later on. The main result of this section is as follows:

Theorem 6.1. Let Ω be a convex polygonal domain in $\mathbb{R}^2$. Suppose that $u \in C^1(\bar{\Omega}) \cap C_0(\bar{\Omega})$ and $u_h \in S_0^h(\Omega)$ satisfy

$$(6.1) \qquad (\nabla u-\nabla u_h, \nabla\phi) = 0, \quad \forall\ \phi \in S_0^h(\Omega).$$

Then there exists a constant C independent of u, u_h and h such that for h sufficiently small

$$(6.2) \qquad \|u-u_h\|_{C(\bar{\Omega})} \le C\, h\, (\ln 1/h) \inf_{\chi \in S_0^h} \|\nabla(u-\chi)\|_{L^\infty(\Omega)} .$$

Remark 6.1. If $u \in C^2(\bar{\Omega}) \cap \overset{\circ}{C}(\bar{\Omega})$ then using the approximation result (3.7) together with (6.2) it follows immediately that

$$(6.3) \qquad \|u-u_h\|_{C(\bar{\Omega})} \le Ch^2(\ln 1/h)\, |u|_{C^2(\bar{\Omega})} ,$$

which is almost optimal from the point of view of approximation theory. Let us also remark that $C^2(\bar{\Omega})$ may be replaced by $W^{2,\infty}(\Omega)$ in (6.3).

Method 1. Estimates using the Greens' function (an outline).

Let x be any point of Ω and denote by $G(x,y)$ the Greens Function for $-\Delta$, i.e. $G(x,y)$ satisfies

$$(6.4) \qquad -\Delta G_y(x,y) = \delta(x-y) \quad \text{in } \Omega$$

and

$$(6.5) \qquad G(x,y) = 0, \ \forall\, y \in \partial\Omega,$$

where $\delta(x-y)$ is the Dirac measure concentrated at x. In term of G we have the representation

$$(6.6) \qquad (u-u_h)\,(x) = \int_\Omega \nabla(u(y)-u_h(y))\cdot\nabla G(x,y)dy$$

Now let $G_h(x,y) \in \overset{o}{S}_h(\Omega)$ be the finite element approximation to the Greens' function defined by

$$(6.7) \qquad \int_\Omega \nabla\, G_h(x,y)\ \nabla\ \chi(y)dy = \chi(x) \quad \forall\ \chi \in \overset{o}{S}_h$$

or equivalently,

$$(6.8) \qquad \int_\Omega \nabla(G(x,y) - G_h(x,y))\cdot\nabla\chi(y)dy = 0 \quad \forall\ \chi \in S_o^h(\Omega)\ .$$

It follows from (6.1) and (6.8) that

$$(6.9) \qquad (u-u_h)(x) = \int_\Omega \nabla(u(-y)-u_h(y)\cdot\nabla(G(x,y-G_h(x,y))dy$$

$$= \int_\Omega \nabla(u(y)-\chi(y))\cdot\nabla(G(x,y)-G_h(x,y))dy.$$

This may be estimated by

$$(6.10) \qquad |(u-u_h)(x)| \le \|\nabla(u-\chi)\|_{L\infty(\Omega)}\, \|\nabla(G(x,\cdot)-G_h(x,\cdot))\|_{L_1(\Omega)}\ .$$

The estimate (6.2) will follow from (6.10) if we can show that

$$\|\nabla(G(x,\cdot)-G_h(x,\cdot)\|_{L^1(\Omega)} \le C\, h(\ln 1/h). \tag{6.11}$$

An estimate of this type is plausible since the second derivatives of $G(x,y)$ "almost" belongs to $L^1(\Omega)$.

Obtaining the estimate (6.11) is the heart of the matter.

This approach was taken by Scott [15] who first proved (6.11) for the finite element method with quasi-uniform triangulations on a smooth domain. A similar approach for finite difference methods had been used by Bramble [1], [2] and Bramble-Hubbard [3]. Scott estimates the left hand side of (6.11) in two parts. i.e.

$$\begin{aligned}\|\nabla(G(x,\cdot)-G_h(x,\cdot)\|_{L^1(\Omega)} &\le \|\nabla(G(x,\cdot)-G_h(x,\cdot))\|_{L^1(B'(x,C^*h))} \\ &\quad + \|\nabla(G(x,\cdot)-G_h(x,\cdot))\|_{L^1(\Omega/B'(x,C^*h))},\end{aligned} \tag{6.12}$$

where $B'(x,C^*h)$ is the intersection of a ball of radius C^*h (C^* chosen appropriately large) with Ω. Note that G is harmonic outside of $B'(x,C^*h)$ which contains the singular part of G. Let us only briefly discuss the second term on the right of (6.12). This may be converted to a weighted L^2 integral by using the Cauchy-Schwarz inequality, i.e.

$$\begin{aligned}\int_{\Omega/B'} |\nabla(G(x,y)-G_h(x,y))|dy &\le \int_{\Omega/B'} r^{-1} r|\nabla(G-G_h)|dy \\ &\le \Big(\int_{\Omega/B'} r^2|\nabla(G-G_h)|^2dy\Big)^{1/2} \Big(\int_{\Omega/B'} r^{-2}dy\Big)^{1/2} \\ &\le C\Big(\ln \frac{1}{h}\Big)^{1/2} \Big(\int_{\Omega/B'} r^2|\nabla G-G_h|^2dy\Big)^{1/2}\end{aligned}$$

This last weighted L^2 integral can be estimated using the ideas used in the proof of Theorem 5.1, where essentially the "cutoff" function w is replaced by the non-vanishing weight r^2 on Ω/B' and $r = |x-y|$.

<u>Method 2. Estimates using Weighted Norms.</u> This method was first used by Natterer [7] and then generalized and put into its present form by Nitsche [8]. Let x_0 be the point in $\bar{\Omega}$ where $\|u_h\|_{C(\bar{\Omega})} = |u_h(x_0)|$, and let $x_0 \in \bar{T}$ for some triangle $T \in \pi_n$. Then using the inverse property (3.18)

$$(6.14) \quad \|u_h\|_{C(\bar{\Omega})} \leq Ch^{-1}\|u_h\|_{L^2(T)} \leq Ch^{-1+\alpha}(\int_\Omega (r^2+C^*h^2)^{-\alpha}\, u_h^2 dx)^{1/2} ,$$

for α a constant to be appropriately chosen and where $r = |x-x_0|$. Notice that

$$(6.15) \qquad h^{-1+\alpha}(\int_\Omega (r^2+C^*h^2)^{-\alpha}\, u^2 dx)^{1/2} \leq C\|u\|_{C(\bar{\Omega})} \quad \text{for} \quad \alpha > 1.$$

The last two inequalities say that the weighted L^2 norm shown above is equivalent to the maximum norm on the subspace $S_0^h(\Omega)$ provided $\alpha > 1$. For C^* a positive constant to be chosen, let

$$(6.16) \qquad \|u\|_\alpha = (\int_\Omega (r^2+C^*h^2)^{-\alpha}\, u^2 dx) .$$

Nitsche [8] proves a general result for domains with a smooth boundary which when restricted to 2 dimensions takes the form

$$(6.17) \qquad \|u_h\|_{\alpha+1} + \|\nabla u_h\|_\alpha \leq C(\|u\|_{\alpha+1} + \|\nabla u\|_\alpha)$$

for $\alpha \in (1,2)$. This says that the finite element projection is bounded in the above weighted norms and the appropriate error estimates follow from (6.14) -- (6.17) and a weighted norm approximation result.

Let us finish the discussion of this method by indicating how the proof of (6.17) starts.

Let $\mu = (r^2 + C^*h^2)$, then for the second term on the left of (6.17).

$$\begin{aligned}(6.18) \quad \|\nabla u_h\|_\alpha^2 &= (\mu^{-\alpha}\nabla u_h, \nabla u_h) = (\nabla u_h, \nabla(\mu^{-\alpha}u_h)) - (\nabla u_h, u_h \nabla\mu^{-\alpha}) \\ &= (\nabla u_h, \nabla(\mu^{-\alpha}u_h)) - \tfrac{1}{2}(\nabla(u_h)^2 \cdot \nabla\mu^{-\alpha}) \\ &= (\nabla u_h, \nabla(\mu^{-\alpha}u_h)) + \tfrac{1}{2}(u_h{}^2, \nabla\mu^{-\alpha}) \\ &= I_1 + I_2\end{aligned}$$

For I_2, we have that since $\nabla\mu^{-\alpha} \leq C\,\mu^{-\alpha-1}$

$$(6.19) \qquad |I_2| \leq C\,\|u_h\|^2_{\alpha+1}$$

In view of (6.1) we have for I_1 the identity

$$(6.20) \qquad (\nabla u_h, \nabla(\mu^{-\alpha}u_h)) = (\nabla u_h, \nabla(\mu^{-\alpha}u_h - \chi)) - (\nabla u, \nabla(\mu^{-\alpha}u_h - \chi) + (\nabla u, \nabla\mu^{-\alpha}u_h)$$

for any $\chi \in S_0^h(\Omega)$.

For the first term on the right of (6.20)

$$|(\nabla u_h, \nabla(\mu^{-\alpha}u_h - \chi))| = |(\mu^{-\alpha/2}, \nabla u_h, \mu^{\alpha/2}\nabla(\mu^{-\alpha}u_h - \chi))| \leq \|\nabla u_h\|_\alpha \, \|\nabla(\mu^{-\alpha}u_h - \chi)\|_{-\alpha} \; .$$

Notice that $\mu^{-\alpha}u_h$ is of the form of a smooth function multiplied by a function in the subspace. One can prove a superapproximation property of the subspace in weighted norms directly analogous to Theorem 3.3. We shall not give further details here but instead refer the reader to Nitsche [8] where a generalization of Theorem 5.1 is proved on smooth domains. Let us just remark that the first step (6.18) in the proof is directly analogous to (5.13)in the proof of local estimates with the cut off function w now replaced by the weight $\mu^{-\alpha}$.

We now turn to a complete proof of theorem 6.1 (see Schatz-Wahlbin [13]).

Step 1: We shall first reduce the proof of (6.2) to the problem of finding an estimate for the Galerkin projection of a function v which may be thought of as a smoothed out and renormalized Greens function. This will be done by first establishing a simple relationship between the maximum norm of the error and the L^2 norm on some triangle T. The L^2 norm of the error will be estimated by a duality argument where v is the solution of the dual problem.

Let $x_0 \in \bar{\Omega}$ be such that

$$(6.21) \qquad \|u - u_h\|_{C(\bar{\Omega})} = |(u - u_h)(x_0)|.$$

Now $x_0 \in \bar{T}$ for some triangle T. Let u_I be the linear interpolant of u on $\bar{T}$, then successively using the triangle inequality, the inverse property (3.20), and the triangle inequality once more it follows that

$$|(u-u_h)(x_0)| \leq |(u(x_0)-u_I)(x_0)| + |(u_I-u_h)(x_0)|$$
$$\leq |u(x_0)-u_I(x_0)| + Ch^{-1}\|u_I-u_h\|_{L^2(T)}$$
$$\leq |u(x_0)-u_I(x_0)| + Ch^{-1}\|u-u_I\|_{L^2(T)} + Ch^{-1}\|u-u_h\|_{L^2(T)}.$$

Since the $\text{mes}(T) \leq Ch^2$, $\|(u-u_I)\|_{L^2(T)} \leq Ch\|u-u_I\|_{C(\bar{T})}$ and hence

$$\|u-u_h\|_{C(\bar{\Omega})} \leq C(\|u-u_I\|_{C(\bar{T})} + h^{-1}\|u-u_h\|_{L^2(T)}).$$

Using the approximation result (3.4), it then follows that

$$(6.22)\quad \|u-u_h\|_{C(\bar{\Omega})} \leq C\, h\|\nabla u\|_{L_\infty(\bar{\Omega})} + Ch^{-1}\|u-u_h\|_{L^2(T)}.$$

The second term on the right will be estimated by a duality argument. Now

$$(6.23)\quad \|u-u_h\|_{L^2(T)} = (u-u_h,\psi),$$

where ψ is defined by

$$\psi(x) = \begin{cases} (u-u_h)(x)/\|u-u_h\|_{L^2(T)} & \text{for } x \in \bar{T} \\ 0 & \text{for } x \in \bar{\Omega} - \bar{T} \end{cases}$$

Notice that $\psi \in L^2(\Omega)$ with $\|\psi\|_{L^2(\Omega)} = \|\psi\|_{L^2(T)} = 1$. Let $v \in W_0^{1,2}(\Omega)$ be the unique solution of the boundary value problem

$$(6.24)\quad (\nabla v,\nabla\phi) = (\psi,\phi) \quad , \quad \forall\ \phi \in W_0^{1,2}(\Omega).$$

From Theorem 2.1 we have that $v \in W^{2,2}(\Omega) \cap W_0^{1,2}(\Omega)$ and

(6.24a) $$\|v\|_{2,2,\Omega} \leq C$$

Notice that v is harmonic on $\Omega - \bar{T}$ and may be thought of as a smoothed out (and renormalized) Green's function. Using (6.24) with $\phi = u-u_h$ in (6.23) we obtain

(6.25) $$\|u-u_h\|_{L^2(T)} = (\nabla(u-u_h),\nabla v)$$

Now let $v_h \in S_0^h(\Omega)$ be the finite element approximation of V, defined by

(6.26) $$(\nabla(v-v_h) ,\nabla\phi) = 0 \quad \forall \phi \in S_0^h .$$

Then

(6.27) $$\|u-u_h\|_{L^2(T)} = (\nabla(u-u_h), \nabla(v-v_h)) = (\nabla u,\nabla(v-v_h))$$
$$\leq \|\nabla u\|_{L_\infty(\Omega)} \|\nabla(v-v_h)\|_{L^1(\Omega)}$$

Combinign (6.27) and (6.22) we arrive at

(6.28) $$\|u-u_h\|_{L_\infty} \leq Ch\|\nabla u\|_{L_\infty(\Omega)} (1 + h^{-2}\|\nabla(v-v_h)\|_{L^1(\Omega)})$$

and in view of the statements immediately after (5.12), the proof of Theorem 6.1 will be complete once the following has been proved:

Step 2:

Lemma 6.1. There exists a constant C independent of h, v, v_h, and x_0. Such that

(6.29) $$\|\nabla(v-v_h)\|_{L^1(\Omega)} \leq Ch^2 \ln(1/h).$$

Proof: Let us first remark that the estimate (6.29) is analogous to (6.11), except that the function v treated here is normalized in a different way. For technical reasons we shall need to choose a fixed positive constant $\hat{C}$ (independent of h and x_0) in the following way: let C^* be as in theorem 5.1 and C^{**} be such that $\max_{T\in\pi_h} \text{Diam}(T) \leq \frac{C^{**}}{2}h$. We then choose $\hat{C}^* = \max(8C^*, 8C^{**})$. Having fixed $\hat{C}$ we turn to proving (6.29). Now let $\hat{\Omega} = B(x_0, \hat{C}h) \cap \Omega$ then (analogous to (6.12))

$$(6.30) \quad \|\nabla(v-v_h)\|_{L^1(\Omega)} = \|\nabla(v-v_h)\|_{L^1(\hat{\Omega})} + \|\nabla(v-v_h)\|_{L^1(\Omega/\hat{\Omega})}$$
$$= J_1 + J_2$$

Using the Cauchy-Schwarz inequality, Theorem 4.2 and (6.24)

$$(6.31) \quad |J_1| \leq Ch\|\nabla(v-v_h)\|_{L^2(\Omega)} \leq Ch\|\nabla(v-v_h)\|_{L^2(\Omega)}$$
$$\leq Ch^2|v|_{2,2,\Omega} \leq Ch^2$$

In order to estimate J_2 we decompose the domain $\Omega/\hat{\Omega}$ as follows: Let A_j be the annular regions

$$A_j = \{x : 2^{-j-1} \leq |x-x_0| \leq 2^{-j}\}, \ j=0,\ 1,\ 2,\ \ldots\ \text{etc.}$$

Set

$$\Omega_j = A_j \cap \Omega$$

and

$$d_j = 2^{-j}$$

Without loss of generality we may assume that $\text{diam}(\Omega) \leq 1$, then $\bar{\Omega} \subseteq \bigcup_{j=0}^{\infty} \Omega_j$. Let J be defined by

$$2^{-J-1} \leq \hat{C}h \leq 2^{-J}$$

Notice that $J = O(\ln \frac{1}{h})$ and $\overline{\Omega/\hat{\Omega}} \subseteq \bigcup_{j=0}^{J} \Omega_j$. We shall need some further notation. Let

$$\Omega_j' = \Omega_{j-1} \cup \Omega_j \cup \Omega_{j+1}$$

$$\Omega_j'' = \Omega_{j-1}' \cup \Omega_j' \cup \Omega_{j+1}' \qquad \text{etc.}$$

Now using the Cauchy-Schwarz inequality

$$J_2 \leq \sum_{j=0}^{J} \|\nabla(v-v_h)\|_{L^2(\Omega_j)} \leq \sum_{j=0}^{J} d_j \|\nabla(v-v_h)\|_{L^2(\Omega_j)}$$

Note that $d_j \geq C^*h$ in this sum. Applying Corollary 5.1 of Theorem 5.1 to the regions Ω_j and Ω_j' respectively it follows from (5.1), with χ chosen to be the interpolant of V that

$$(6.32) \qquad J_2 \leq C\Big(\sum_{j=0}^{J} d_j h |v|_{2,2,\Omega_j'} + \sum_{j=0}^{J} \|v-v_h\|_{L^2(\Omega_j')}\Big)$$

For the second term on the right of (6.32)

$$(6.33) \qquad \sum_{j=0}^{J} \|v-v_h\|_{L^2(\Omega_j')} \leq (\ell n\ 1/h)^{1/2} \Big(\sum_{j=0}^{J} \|v-v_h\|^2_{L^2(\Omega_j')}\Big)^{1/2}$$

$$\leq (\ell n\ 1/h)^{1/2} \|v-v_h\|_{L^2(\Omega)}$$

$$\leq Ch^2(\ell n\ 1/h)^{1/2} |v|_{2,2,\Omega} \leq Ch^2(\ell n\ 1/h)^{1/2},$$

where we have used the Cauchy-Schwarz inequality, Theorem 4.3 and (6.24a). Combining (6.33), (6.32) and (6.31) with (6.30) we arrive at

$$(6.34) \qquad \|\nabla(v-v_h)\|_{L^1(\Omega)} \leq C(h^2(\ln 1/h)^{1/2} + h \sum_{j=0}^{J} d_j\ |v|_{2,2,\Omega_j'}).$$

Because of our choice of $\hat{C}$, the function V is harmonic inside Ω_j''' $j = 0, \ldots. J$ and the following local estimate holds.

$$(6.35) \qquad |v|_{2,2,\Omega_j'} \leq C\ d_j^{-1} \|\nabla v\|_{L^2(\Omega_j''')}.$$

Granting this estimate for a moment, let us complete the proof of (6.29). Now for the second term on the right in (6.34)

$$(6.36) \qquad h \sum_{j=0}^{J} d_j \, |V|_{2,2,\Omega_j'} \le Ch \sum_{j=0}^{J} \|\nabla V\|_{L^2(\Omega_j'')}$$

$$\le Ch(\ell n\ 1/h)^{1/2} \left(\sum_{j=0}^{J} \|\nabla v\|^2_{L^2(\Omega_j'')} \right)^{1/2}$$

$$\le Ch(\ell n\ 1/h)^{1/2} \|\nabla v\|_{L^2(\Omega)}$$

In order to estimate this last term we use

$$\|\nabla v\|^2_{L^2(\Omega)} = (\nabla v, \nabla v) = (\psi, v) = (\psi, v - v_h) + (\psi, v_h)$$

$$\le \|\psi\|_{L^2(\Omega)} \|v - v_h\|_{L^2(\Omega)} + \|\psi\|_{L^1(T)} \|v_h\|_{C(\bar{T})}$$

Using (6.24), Theorem 4.3, the estimate $\|\psi\|_{L^1(T)} \le Ch\|\psi\|_{L^2(T)} = Ch$, and the Sobolev inequality (3.23)

$$\|\nabla v\|^2_{L^2(\Omega)} \le Ch(\|\nabla v\|_{L^2(\Omega)} + (\ell n\ 1/h)^{1/2} \|\nabla v_h\|_{L^2(\Omega)}$$

$$\le C\,h(1 + (\ell n\ \tfrac{1}{h})^{1/2}) \|\nabla v\|_{L^2(\Omega)}$$

or

$$\|\nabla v\|_{L^2(\Omega)} \le Ch(\ell n\ 1/h)^{1/2}$$

This estimate combined with (6.36) and (6.34) proves (6.29). We now turn to the proof of (6.35).

Let $w \in C_0^\infty(\Omega_j'')$, $w \ge 0$, $w \equiv 1$ on Ω_j and

$$|D^\alpha w| \le Cd^{-|\alpha|} \quad , \quad |\alpha| = 1,2$$

We consider two cases. If Ω_j'' does not intersect $\partial\Omega$ then for any constant M, $w(v-M)\varepsilon W^{2,2}(\Omega) \cap W_0^{1,2}(\Omega)$ and using the apriore estimate (2.3) and the fact that $\Delta(v-M)= 0$ on Ω_j''

$$(3.37) \qquad |v|_{2,2,\Omega_j'} = |v-M|_{2,2,\Omega_j'} \leq |w(v-M)|_{2,2,\Omega} \leq \|\Delta w(v-M)\|_{L^2(\Omega)}$$

$$= \|w\Delta(v-M) + \nabla w\cdot\nabla(v-M) + (v-M)\Delta w\|_{L^2(\Omega)}$$

$$\leq C(d_j^{-1} \|\nabla v\|_{L^2(\Omega_j'')} + d_j^{-2} \|v-M\|_{L^2(\Omega_j'')}).$$

Choosing $M = \int_{\Omega_j''} v\,dx$ it easily follows from a Poincare inequality that

$$\|v-M\|_{L^2(\Omega_j'')} \leq d_j \|\nabla(v-M)\|_{L^2(\Omega_j'')} = d_j \|\nabla v\|_{L^2(\Omega_j'')}.$$

This combined with (3.37) yield (6.35). If $\Omega_j'' \cap \partial\Omega \neq 0$ we take $M = 0$ and from (3.77) obtain

$$|v|_{2,2,\Omega_j'} \leq C(d_j^{-1} \|\nabla v\|_{L^2(\Omega_j'')} + d_j^{-2} \|v\|_{L^2(\Omega_j'')})$$

Extending v to be zero outside of Ω, then

$$\|v\|_{L^2(\Omega_j'')} = \|v\|_{L^2(A_j'')} \leq \|v\|_{L^2(A_j''')}.$$

Now A_j''' contains a set of measure $\geq Cd^2$ on which v vanishes and a Poincare inequality yields

$$\|v\|_{L^2(A_j''')} \leq d_j\|\nabla v\|_{L^2(A_j''')} = d_j\|\nabla v\|_{L^2(\Omega_j''')}.$$

Taken together these last three inequalities yield (6.35) which completes the proof.

Let us end this section with a discussion of other results relating to maximum norm estimates.

First let Ω be a bounded domain in $\mathbb{R}^N$, $N \geq 2$ (for the case $N = 1$ see Douglis-Dupont-Wahlbin [5]) with a smooth boundary. Let $S_0^{h,r}$ denote the space of piecewise polynomials of degree $r-1$, $r \geq 2$ on a quasi-uniform triangulation of Ω which vanish on $\partial\Omega$ (i.e. the boundary is assumed to be fitted exactly). Using Method 2 Nitsche proved

$$\|u-u_h\|_{C(\bar{\Omega})} \leq C\, h(\ln \tfrac{1}{h})^{\bar{r}} \inf_{\chi \in S_0^{h,r}} \|\nabla(u-\chi)\|_{L_\infty(\bar{\Omega})},$$

where $\bar{r} = 1$ if $r = 2$, and $\bar{r} = 0$ if $r \geq 3$. Using isoparametric elements Schatz and Wahlbin [ ] showed

$$\|u-u_h\|_{C(\bar{\Omega})} \leq C(\ln \tfrac{1}{h})^{\bar{r}} \inf_{\chi \in S_0^{h,r}} \|u-\chi\|_{C(\bar{\Omega})}, \tag{3.38}$$

which says the Galerkin finite element projection is maximum norm stable for $r \geq 3$ and almost maximum norm stable for $r = 2$. In Schatz [11] it was shown that (3.38) also holds on plane polygonal domains (both convex and non-convex). In the case that Ω is a plane convex polygonal domain Scott and Rannacher [14] have proved

$$\|\nabla(u-u_h)\|_{L_\infty} \leq C \inf_{\chi \in S_0^h} \|\nabla(u-\chi)\|_{L_\infty},$$

which shows that the gradients of the Galerkin finite element projection are stable in L_∞ for piecewise linear functions. For $S_0^{h,r}$ $r \geq 3$ this result was known previously, and for S_0^h this nicely removes a logarithm from known estimates. For a rather complete bibliography on maximum norm estimates the reader is referred to Nitsche [9].

BIBLIOGRAPHY

[1] J. H. Bramble, "On the convergence of difference approximations to week solutions of Dirichlet's problem", Num. Math. 13, 1969, pp. 101-111.

[2] J. H. Bramble, "On the convergence of difference approximations for second order uniformly elliptic operators", SIAM-AMS Proceedings 2, 1970, pp. 201-210.

[3] J. H. Bramble and B. E. Hubbard, "A priore bounds on the descretization error in the numerical solution of the Dirichlet problem", Contributions to Differential Equations 2, 1963, pp. 229-252.

[4] Ph. Ciarlet, The Finite Element Method for Elliptic Problems, North-Holland, Amsterdam, 1978.

[5] J. Douglas, Jr., T. Dupont and L. B. Wahlbin, "Optimal L_∞ error estimates for Galerkin approximations of solutions of two point boundary value problems", Math. Comp., V.29, 1975, pp. 475-483.

[6] P. Grisvard, "Boundary value problems on non-smooth domains, Part I", Lecture Notes, University of Maryland, 1979.

[7] F. Natterer, "Uber die punktweize konvergenz finiter elemente", Math. Z., V. 149, 1976, pp. 69-77.

[8] J. A. Nitsche, "L_∞-convergence of finite element approximations, mathematical aspects of finite element methods", Lecture Notes in Mathematics, Vol. 606, Springer-Verlag, New York, 1977, pp.261-274.

[9] J. A. Nitsche, "L_∞ error analysis for finite elements", The Mathematics of Finite Elements and Applications III, J. R. Whiteman ed., Academic Press, London, 1979, pp. 173-186.

[10] A. Nitsche and A. H. Schatz, "Interior estimates for Ritz-Galerkin methods", Math. Comp., V. 28, 1974, pp. 937-958.

[11] A. H. Schatz, "A weak discrete maximum principle and stability of the finite element method in L_∞ on plane polygonal domains", Math. Comp., V. 34, 1980, pp. 77-91.

[12] A. H. Schatz and L. B. Wahlbin, "On the quasi-optimality in L_∞ of the $\overset{\circ}{H}{}^1$-projection into finite element spaces", Math. Comp. V. 38, 1982, pp. 1-22.

[13] A. H. Schatz and L. B. Wahlbin, "Maximum norm estimates for the finite element method on plane polygonal domains, Part I", Math. Comp. V. 32, 1978, pp. 73-109.

[14] R. Rannacher and R. Scott, "Some optimal error estimates for piecewise linear approximations", Math. Comp. V. 38, 1982, p. 437-446.

[15] R. Scott, "Optimal L^∞ estimates for the finite element method on irregular meshes", Math. Comp. V. 30, 1976, pp. 681-697.

[16] R. E. Showalter, Variational Theory and Approximation of Boundary Value Problems, Lecture Notes, this issue.

VARIATIONAL THEORY AND APPROXIMATION OF BOUNDARY VALUE PROBLEMS

R.E. Showalter
Department of Mathematics
University of Texas
Austin, Texas 78712

I. Introduction

Suppose we wish to construct an approximation of the solution u of an elliptic boundary value problem which is written in the operator form $Lu = f$. If this is an equation of functionals acting on test functions v belonging to a space V, this means $Lu(v) = f(v)$, $v \in V$. Galerkin's scheme is to replace the usually infinite dimensional space V by a finite dimensional subspace S of V. Thus we seek a $U \in S$ such that $LU(v) = f(v)$, $v \in S$. When the dimension of S is m, this latter problem reduces to a linear algebra problem of m equations in m unknowns. To implement this scheme numerically the space S should be chosen so that the calculations are convenient and that S contains some elements which are very close to u. Both of these conditions are remarkably well achieved by finite element spaces of piecewise polynomial functions: the resulting linear algebra problem for the approximate solution U is given with a sparse matrix, and approximation theory provides precise estimates of the rate of best approximation by the space S. The variational formulation of the problem provides a solid mathematical theory for both the original and the approximate problems and, moreover, implies directly that the Galerkin approximation U actually achieves the optimal convergence rate from S. In summary, the basic steps in the development of a finite element Galerkin method are the:

(i) variational formulation of a well-posed problem,
(ii) construction of appropriate finite element subspaces,
(iii) approximation theory of distance from S to u, and
(iv) numerical analysis of the large linear algebra problems.

Our plan in the following is to develop the first of these topics and provide a variety of examples of boundary value problems that arise in applications. The techniques chosen for presentation are among those which extend immediately to more general (nonlinear) and important situations, and we discuss these general principles in the least technical but still useful contexts. The remaining three topics will be exemplified in the most simple case (of dimension = 1) in order to provide a rather complete overview for orientation and motivation.

We begin by considering two classical boundary value problems. Let $H \equiv L^2(a,b)$, the space of (equivalence classes of) square-summable functions on the real interval (a,b), and let $c \in \mathbb{R}$. The Dirichlet problem is to find

$$u \in H: \ -u'' + cu = F \quad \text{in} \quad H, \ u(a) = u(b) = 0 ,$$

and the Neumann problem is to find

$$u \in H: \ -u'' + cu = F \quad \text{in} \quad H, \ u'(a) = u'(b) = 0 .$$

An implicit requirement of these classical formulations is that $u'' \in H$. This can be relaxed: multiply the equation by $v \in H$ and integrate. If also $v' \in H$ we obtain the following by an integration-by-parts. Let $V_0 \equiv \{v \in H: v' \in H \text{ and } v(a) = v(b) = 0\}$; a solution of the Dirichlet problem satisfies

$$u \in V_0 \quad \text{and} \quad \int_a^b (u'v' + cuv)\,dx = \int_a^b Fv\,dx, \ v \in V_0.$$

Similarly, a solution of the Neumann problem satisfies

$$u \in V_1 \quad \text{and} \quad \int_a^b (u'v' + cuv)\,dx = \int_a^b Fv\,dx, \ v \in V_1$$

where $V_1 \equiv \{v \in H: v' \in H\}$. These are the corresponding weak formulations of the respective problems. We shall see directly that they are actually equivalent to their classical formulations. Moreover we see already the primary ingredients of the variational theory:

(a) functionals. Each function, e.g., $F \in H$, is identified with a functional, $\tilde{F}: H \to \mathbb{R}$, defined by $\tilde{F}(v) \equiv \int_a^b Fv\,dx$, $v \in H$. This identification is achieved by way of the L^2 scalar product.

For a pair $u \in V_1$, $v \in V_0$ an integration-by-parts shows $\tilde{u}'(v) = -\tilde{u}(v')$. Thus, for this identification of functions with functionals to be consistent with the usual differentation of functions, it is necessary to define the generalized derivative of a functional f by $\partial f(v) \equiv -f(v')$, $v \in V_0$.

(b) function spaces. From $L^2(a,b)$ and the generalized derivative ∂ we construct the Sobolev space $H^1(a,b) \equiv \{v \in L^2(a,b) : \partial v \in L^2(a,b)\}$. We shall see directly that $H^1(a,b)$ is a Hilbert space with scalar product

$$(u,v)_{H^1} \equiv \int_a^b (\partial u \partial v + uv)\,dx$$

and norm $\|u\|_{H^1} = (u,u)^{\frac{1}{2}}_{H^1}$, and each of its members is absolutely continuous, hence, $u(x) - u(y) = \int_y^x \partial u$ for $u \in H^1(a,b)$, $a < y < x < b$. From this follows immediately the estimate

$$|u(x) - u(y)| \leq |x-y|^{\frac{1}{2}} \|\partial u\|_{L^2} \leq (b-a)^{\frac{1}{2}} \|\partial u\|_{L^2}, \qquad x,y \in (a,b) \qquad [1.1]$$

for $u \in H^1(a,b)$. If, in addition, $u(a) = 0$ we obtain

$$\|u\|_{L^\infty} = \max_{a \leq x \leq b} \{|u(x)|\} \leq (b-a)^{\frac{1}{2}} \|\partial u\|_{L^2}, \qquad [1.2]$$

$$\|u\|_{L^2} \leq ((b-a)/\sqrt{2}) \|\partial u\|_{L^2}. \qquad [1.3]$$

In fact, there is a constant $k > 0$ such that

$$\|u\|_{L^\infty} \leq k \|u\|_{H^1}, \qquad u \in H^1(a,b) ; \qquad [1.4]$$

this shows that pointwise evaluations, such as the trace functionals $\gamma_a(v) \equiv v(a)$, $v \in H^1$, are continuous. It follows that the subspace $H_0^1(a,b) \equiv \{v \in H^1(a,b) : v(a) = v(b) = 0\}$ is closed, hence is a Hilbert space.

(c) forms. Each of our weak formulations is phrased as

$$u \in V : a(u,v) = f(v), \quad v \in V \qquad [1.5]$$

where V is an appropriate Hilbert space (either H_0^1 or H^1), $f = \tilde{F}$ is a continuous function on V, and $a(\cdot,\cdot)$ is the bilinear form on V defined by

$$a(u,v) = \int_a^b (\partial u \partial v + cuv)\,dx, \quad u,v \in V .$$

This form is bounded or continuous on V:

$$|a(u,v)| \leq K\|u\|_V \|v\|_V , \quad u,v \in V . \qquad [1.6]$$

Moreover, it is V-coercive, i.e., there is a $c_0 > 0$ for which

$$|a(v,v)| \geq c_0 \|v\|_V^2 , \quad v \in V \qquad [1.7]$$

in the case of $V = H^1(a,b)$ if (and only if!) $c > 0$ and in the case of $V = H_0^1(a,b)$ for any $c > -2/(b-a)^2$. (This last inequality follows from [1.3] but is it not the optimal constant.) We shall see that the weak formulation constitutes a well-posed problem whenever the bilinear form is bounded and coercive.

Next we consider a general Galerkin approximation of the weak formulation [1.5] which we assume is well-posed as above. Thus, let S be any closed subspace of V; the same theory applied with S instead of V shows there exists exactly one

$$U \in S : a(U,v) = f(v) , \quad v \in S . \qquad [1.8]$$

A natural choice of S is suggested by any basis $\{v_1, v_2, v_3, \ldots\}$ of the (separable) space V. For each integer $m \geq 1$ let V_m be the linear span of the set $\{v_1, v_2, \ldots, v_m\}$. Taking $S = V_m$ we obtain an equivalent problem of determining the coefficients $(u_1, u_2, \ldots, u_m) \in \mathbb{R}^m$ in the expansion of the solution of [1.8] as $U = \sum_{j=1}^{m} u_j v_j$. Thus [1.8] is equivalent to the $m \times m$ linear system

$$\sum_{i=1}^{m} a(v_i,v_j)u_i = f(v_j) \ , \quad 1 \le j \le m \ . \qquad [1.9]$$

In general the problem [1.9] is large and possibly difficult to solve numerically; we know only that the matrix is symmetric and invertible. Since $\{v_j\}$ is a basis for V it follows $\lim_{m\to\infty} \inf\{\|u-v\|_V : v \in V_m\} = 0$, but the convergence may be slow. As a computational scheme this might not be a good situation.

There is an apparent optimal choice above: suppose $a(v_i,v_j) = 0$ for all $i \ne j$. Then the matrix is diagonal, [1.9] is trivial to solve and, moreover, each component u_j is independent of $m \ge j$. The only difficulty is the possible slowness of convergence. Such a basis is orthogonal, and the corresponding eigenvalue problem of finding such a basis is more difficult than resolving [1.5]!

However we can directly construct a subspace of $V = H^1(a,b)$ for which the matrix $A = (a(v_i,v_j))$ is sparse. Partition the interval (a,b) with points $a = x_1 < x_2 < \ldots x_m = b$ and let $h \equiv \max\{x_{j+1}-x_j : 1 \le j \le m-1\}$ denote the mesh size. Define v_k to be that continuous function on (a,b) which is affine on each subinterval (x_j,x_{j+1}) and satisfies $v_k(x_j) = 1$ for $j = k$ and $= 0$ for $j \ne k$, $1 \le j, k \le m$. Then the bilinear form satisfies $a(v_i,v_j) = 0$ for $|i-j| \ge 2$ so the matrix in [1.9] is tridiagonal: all entries off the diagonal and its immediate neighbors are zero. Thus the linear algebra problem is easy to solve no matter how large is m.

Next we consider how well functions in $H^1(a,b)$ can be approximated by those of $S_h \equiv$ linear span of $\{v_1,v_2,\ldots,v_m\}$. For each $f \in H^1(a,b)$ it is immediate that the unique piece-wise-affine function agreeing with f at the points x_j is the (Lagrange interpolant) function $f_h \in S_h$ given by $f_h(x) = \sum_{j=1}^{m} f(x_j)v_j(x)$, $a \le x \le b$. Define

$$K \equiv \{v \in H^1(a,b) : v(x_j) = 0 \ , \quad 1 \le j \le m\} \ .$$

<u>Lemma 1</u>. For each $f \in H^1(a,b)$, $f_h - f \in K$ and

$$(\partial f_h, \partial v)_{L^2} = 0 \ , \quad v \in K \ , \qquad \|\partial(f-f_h)\|_{L^2} \le \|\partial f\|_{L^2} \ ,$$

$$\|f-f_h\|_{L^2} \le h\|\partial(f-f_h)\|_{L^2} \ , \qquad \|f-f_h\|_{L^2} \le h\|\partial f\|_{L^2} \ .$$

Proof: The first inclusion is obvious and the orthogonality statement follows by summing

$$\int_{x_j}^{x_{j+1}} \partial f_h \partial v dx = - \int_{x_j}^{x_{j+1}} \partial^2 f_h v dx = 0 .$$

Thus $f - f_h$ is the projection of f onto K with the scalar product $(\partial f, \partial g)_{L^2}$, and the identity $\|\partial f_h\|^2_{L^2} + \|\partial(f_h - f)\|^2_{L^2} = \|\partial f\|^2_{L^2}$ yields the first estimate. The second estimate is obtained by summing (see [1.3])

$$\int_{x_j}^{x_{j+1}} |v|^2 \le h^2 \int_{x_j}^{x_{j+1}} |\partial v|^2 \quad \text{for} \quad v = f - f_h \in K$$

and the third follows the preceding two.

Lemma 2. If, in addition, $\partial^2 f \in L^2(a,b)$, then

$$\|\partial(f - f_h)\|_{L^2} \le h\|\partial^2 f\|_{L^2} , \qquad \|f - f_h\|_{L^2} \le h^2 \|\partial^2 f\|_{L^2} .$$

Proof: A direct computation gives

$$\|\partial(f - f_h)\|^2_{L^2} = -\sum_{j=1}^{m-1} \int_{x_j}^{x_{j+1}} \partial^2(f - f_h)(f - f_h)$$

$$= -(\partial^2 f, f - f_h)_{L^2} \le \|\partial^2 f\|_{L^2} \|f - f_h\|_{L^2} ,$$

so the estimates follow from Lemma 1.

In summary, Lemma 2 shows that the best approximation of a "regular" $f \in H^1(a,b)$ is bounded in terms of mesh size h.

Finally, we estimate the error that results from the replacement of the weak formulation [1.5] by the Galerkin approximation [1.8]. Since this error $u - U$ satisfies $a(u - U, w) = 0$ for all $w \in S$, it follows that $a(u - U, u - U) = a(u - U, u - w)$, $w \in S$. From [1.6] and [1.7] we conclude

$$\|u - U\|_V \leq (K/c_0) \inf\{\|u - w\|_V : w \in S\} .$$

This is the general energy estimate for the error, and for the finite-element space $S = S_h$ we obtain from Lemma 2

$$\|u - U\|_{H^1} \leq (K/c_0)\|\partial^2 u\|_{L^2} h\sqrt{1+h^2} = O(h) . \qquad [1.10]$$

This also bounds the L^2-norm of error, but one can do better. For $u \neq U$ we set $g = (u - U)/\|u - U\|_{L^2}$ and then solve the corresponding problem

$$w \in V : a(w,v) = (g,v)_{L^2} , \qquad v \in V$$

as before. Setting $v = u - U$ then yields

$$\|u - U\|_{L^2} = a(w, u - U) = a(w - z, u - U) , \qquad z \in S$$

and therefore the estimate

$$\|u - U\|_{L^2} \leq K \inf\{\|w - z\|_V : z \in S\}\|u - U\|_V .$$

Specializing to the finite-element space $S = S_h$, we obtain from Lemma 2

$$\|u - U\|_{L^2} \leq (K^2/c_0)\|\partial^2 w\|_{L^2}\|\partial^2 u\|_{L^2} h^2(1+h^2) = O(h^2) . \qquad [1.11]$$

We should note that these estimates depend on the observation that the solution satisfies $\partial^2 u \in L^2$ whenever the data F belongs to L^2. This is trivial for our examples here but is an important result of regularity theory in higher dimensions. A consequence is that $\|\partial^2 u\|_{L^2}$ is bounded by a multiple of $\|F\|_{L^2}$ and, since $\|g\|_{L^2} = 1$, the factor $\|\partial^2 w\|_{L^2}$ is bounded by a constant independent of the solution.

II. Variational Method in Hilbert Space

Our objective is to review certain topics in the elementary theory of Hilbert space which lead directly to abstract variational or weak formulations of boundary value problems. Let V be a linear space over the reals $\mathbb{R}$ and the function $x,y \longmapsto (x,y)$ from $V \times V$ to $\mathbb{R}$ be a scalar product. That is, $(x,x) > 0$ for non-zero $x \in V$, $(x,y) = (y,x)$ for $x,y \in V$, and for each $y \in V$ the function $x \longmapsto (x,y)$ is linear from V to $\mathbb{R}$. For each pair $x,y \in V$ it follows that

$$|(x,y)|^2 \leq (x,x)(y,y) \ . \qquad [2.1]$$

To see this, we note that

$$0 \leq (tx+y,\ tx+y) = t^2(x,x) + 2t(x,y) + (y,y), \qquad t \in \mathbb{R} ,$$

and so the discriminant of the quadratic must be negative. From [2.1] it follows that $\|x\| \equiv (x,x)^{\frac{1}{2}}$, $x \in V$, defines a norm on V: $\|x\| \geq 0$, $\|tx\| = |t|\|x\|$, and $\|x+y\| \leq \|x\|+\|y\|$ for $x,y \in V$ and $t \in \mathbb{R}$. Thus every scalar product induces a norm and corresponding metric $d(x,y) = \|x-y\|$. A sequence $\{x_n\}$ converges to x in V if $\lim\limits_{n\to\infty} \|x_n - x\| = 0$. This is denoted by $\lim\limits_{n\to\infty} x_n = x$. A convergent sequence is always Cauchy: $\lim\limits_{m,n\to\infty} \|x_m - x_n\| = 0$. The space V with norm $\|\cdot\|$ is complete if each Cauchy sequence is convergent in V. A complete scalar product space is a Hilbert space.

Some familiar examples of Hilbert spaces include Euclidean space $\mathbb{R}^m = \{\vec{x} = (x_1,x_2,\ldots,x_m): x_j \in \mathbb{R}\}$ with $(\vec{x},\vec{y}) = \sum_{j=1}^{m} x_j y_j$, the sequence space $\ell^2 = \{\bar{x} = \{x_1,x_2,x_3,\ldots\}: \sum_{j=1}^{\infty} |x_j|^2 < \infty\}$ with $(\vec{x},\vec{y}) = \sum_{j=1}^{\infty} x_j y_j$, and the Lebesgue space $L^2(\Omega) = \{$equivalence classes of measurable $f:\Omega \to \mathbb{R}$: $\int_\Omega |f|^2 d\mu < \infty\}$ with $(f,g) = \int_\Omega f(w)g(w)d\mu$, where Ω,μ is a measure space.

Let V_1 and V_2 be scalar-product spaces with corresponding norms $\|\cdot\|_1, \|\cdot\|_2$. A function $T: V_1 \to V_2$ is continuous at $x \in V_1$ if $\{T(x_n)\}$ converges to $T(x)$ in V_2 whenever $\{x_n\}$ converges to x in V_1.

Proposition. If $T: V_1 \to V_2$ is linear, the following are equivalent:

(a) T is continuous at 0,

(b) T is continuous at every $x \in V_1$,

(c) there is a constant $K \geq 0$ such that $\|Tx\|_2 \leq K\|x\|_1$ for all $x \in V_1$.

Proof: Clearly (c) implies (b) and (b) implies (a). If (c) were false there would be a sequence $\{x_n\}$ in V_1 with $\|Tx_n\|_2 > n\|x_n\|_1$, but then $y_n \equiv \|T(x_n)\|_2^{-1} x_n$ is a sequence which contradicts (a).

We shall denote by $\mathcal{L}(V_1, V_2)$ the set of all continuous linear functions from V_1 to V_2; these are called the bounded linear functions because of (c) above. Additional structure on this set is given as follows.

Proposition. For each $T \in \mathcal{L}(V_1, V_2)$ we have

$$\|T\| \equiv \sup\{\|Tx\|_2 : x \in V_1, \|x\|_1 \leq 1\} = \sup\{\|Tx\|_2 : \|x\|_1 = 1\}$$

$$= \inf\{K > 0 : \|Tx\|_2 \leq K\|x\|_1, \quad x \in V_1\} ,$$

and this gives a norm on $\mathcal{L}(V_1, V_2)$. If V_2 is complete, then $\mathcal{L}(V_1, V_2)$ is complete.

As a consequence it follows that the dual $V' \equiv \mathcal{L}(V, \mathbb{R})$ of any normed linear space V is complete with the dual norm

$$\|f\|_{V'} \equiv \sup\{|f(x)| : x \in V, \|x\|_V \leq 1\}$$

for $f \in V'$.

Hereafter we let V denote a Hilbert space with norm $\|\cdot\|$, scalar product $(\cdot,\cdot)$, and dual space V'. A subset K of V is called closed if each $x_n \in K$ and $\lim x_n = x$ imply $x \in K$. The subset K is convex if $x, y \in K$, $0 \leq t \leq 1$ imply $tx + (1-t)y \in K$. The following minimization principle is fundamental.

Theorem 1. Let K be a closed convex non-empty subset of V and let $f \in V'$. Define $\phi(x) \equiv (\tfrac{1}{2})\|x\|^2 - f(x)$, $x \in V$. Then there exists a unique

$$x \in K : \phi(x) \leq \phi(y), \quad y \in K. \qquad [2.2]$$

Proof: Set $d \equiv \inf\{\phi(y) : y \in K\}$ and choose $x_n \in K$ such that $\lim_{n \to \infty} \phi(x_n) = d$. Then we obtain successively

$$d \leq \phi(\tfrac{1}{2}(x_m + x_n)) = (\tfrac{1}{2})(\phi(x_m) + \phi(x_n)) - (\tfrac{1}{8})\|x_n - x_m\|^2 ,$$

$$(\tfrac{1}{4})\|x_n - x_m\|^2 \leq \phi(x_m) + \phi(x_n) - 2d ,$$

and this last expression converges to zero. Thus $\{x_n\}$ is Cauchy, it converges to some $x \in V$ by completeness, and $x \in K$ since it is closed. Since ϕ is continuous, $\phi(x) = d$ and x is a solution of [2.2]. If x_1 and x_2 are both solutions of [2.2], the last inequality shows $(\tfrac{1}{4})\|x_1 - x_2\| \leq d + d - 2d = 0$, so $x_1 = x_2$.

The solution of the minimization problem [2.2] can be characterized by a variational inequality. For $x, y \in V$ and $t > 0$ we have $(1/t)(\phi(x + t(y - x) - \phi(x)) = (x, y - x) - f(y - x) + (\tfrac{1}{2})t\|y - x\|^2$, so the <u>derivative</u> of ϕ at x in the direction $y - x$ is given by

$$\phi'(x)(y - x) = \lim_{t \to 0} (1/t)(\phi(x + t(y - x)) - \phi(x)) = (x, y - x) - f(y - x) . \qquad [2.3]$$

An easy calculation shows the above equals $\phi(y) - \phi(x) + (x, y) - (\tfrac{1}{2})\|x\|^2 - (\tfrac{1}{2})\|y\|^2$, so [2.1] gives

$$\phi'(x)(y - x) \leq \phi(y) - \phi(x) , \quad x, y \in V . \qquad [2.4]$$

From [2.3] and [2.4] it is immediate that [2.2] is equivalent to

$$x \in K : (x, y - x) \geq f(y - x) , \quad y \in K . \qquad [2.5]$$

The equivalence of [2.2] and [2.5] is merely the fact that the point where a quadratic function takes its minimum is characterized by having a positive derivative in each direction into the set.

As an example, let $x_0 \in V$ and define $f \in V'$ by $f(y) = (x_0, y)$ for $y \in V$. Then $\phi(x) = (\frac{1}{2})\|x - x_0\|^2 - \|x_0\|^2$ so [2.2] means that x is that point of K which is closest to x_0. Recalling that the angle θ between $x - x_0$ and $y - x$ is determined by

$$(x - x_0, y - x) = \cos(\theta)\|x - x_0\|\|y - x\| ,$$

we see [2.5] means x is that point of K for which $-\pi/2 \leq \theta \leq \pi/2$ for every $y \in K$. We define x to be the <u>projection</u> of x_0 on K and denote it by $P_K(x_0)$.

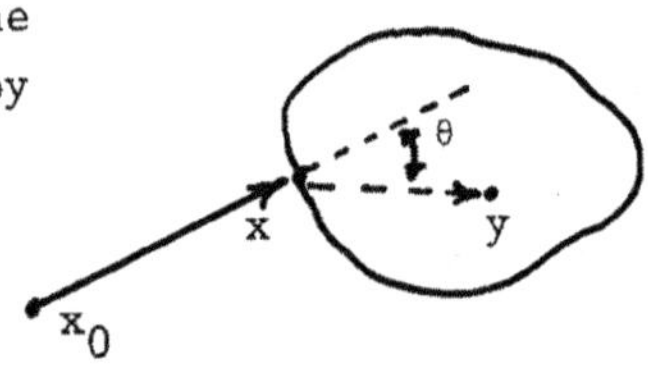

<u>Corollary 1</u>. For each closed convex non-empty subset K of V there is a projection $P_K: V \longrightarrow K$ for which $P_K(x_0)$ is that point of K closest to $x_0 \in V$; it is characterized by

$$P_K(x_0) \in K: (P_K(x_0) - x_0, y - P_K(x_0)) \geq 0 , \quad y \in K .$$

<u>Corollary 2</u>. For each closed subspace K of V and each $x_0 \in K$ there is a unique

$$x \in K: (x - x_0, y) = 0 , \quad y \in K .$$

Two vectors $x, y \in V$ are <u>orthogonal</u> if $(x,y) = 0$, and the <u>orthogonal complement</u> of the set S is $S^\perp \equiv \{x \in V: (x,y) = 0$ for $y \in S\}$. Corollary 2 says each $x_0 \in V$ can be uniquely written in the form $x_0 = x_1 + x_2$ with $x_1 \in K$ and $x_2 \in K^\perp$ whenever K is a closed subspace. We denote this orthogonal decomposition by $V = K \oplus K^\perp$.

The <u>Riesz map</u> $\underset{\sim}{R}$ of V into V' was implicit in the preceding discussion. It is defined by $\underset{\sim}{R}x(y) = (x,y)$ for $x,y \in V$. It is clear that $\|\underset{\sim}{R}x\|_{V'} = \|x\|_V$; Theorem 1 with $K = V$ shows by way of [2.5] that $\underset{\sim}{R}$ is onto V', so $\underset{\sim}{R}$ is an isometric isomorphism of the

Hilbert space V onto its dual V'. Specifically, for each $f \in V'$ there is a unique $x = \underset{\sim}{R}^{-1}(f) \in V$.

Corollary 3. For each $f \in V'$ there is a unique

$$x \in V: (x,y) = f(y) , \quad y \in V . \qquad [2.6]$$

We recognize [2.6] as the weak formulation [1.5] of certain boundary value problems. Specifically, when $V = H_0^1$ or H^1, [2.6] is the Dirichlet or Neumann problem, respectively, with $c = 1$. The trivial but useful generalization is obtained as follows. Let $a: V \times V \to \mathbb{R}$ be bilinear (linear in each variable separately), continuous (cf. [1.6]), symmetric $(a(x,y) = a(y,x), x,y \in V)$ and V-elliptic: there is a $c_0 > 0$ such that

$$a(x,x) \geq c_0 \|x\|^2 , \quad x \in V . \qquad [2.7]$$

Thus, $a(\cdot,\cdot)$ is another scalar product on V which is equivalent to the first: a sequence converges in V with $\|\cdot\|$ if and only if it converges with $a(\cdot,\cdot)^{\frac{1}{2}}$. Thus we may replace $(\cdot,\cdot)$ by $a(\cdot,\cdot)$ above.

Theorem 1a. Let $a(\cdot,\cdot)$ be a bilinear, symmetric, continuous and V-elliptic form on the Hilbert space V, let K be a closed, convex and non-empty subset of V, and let $f \in V'$. Set $\phi(x) = (\frac{1}{2})a(x,x) - f(x)$, $x \in V$. Then there is a unique

$$x \in K: \phi(x) \leq \phi(y) , \quad y \in K . \qquad [2.8]$$

The solution of [2.8] is characterized by

$$x \in K: a(x, y - x) \geq f(y - x) , \quad y \in K . \qquad [2.9]$$

If, in addition, K is a subspace of V, then [2.9] is equivalent to

$$x \in K: a(x,y) = f(y) , \quad y \in K . \qquad [2.10]$$

Now [2.10] is precisely our weak formulation and we see it is the special case of a variational inequality [2.9] which is the characterization of the solution of the minimization problem [2.8]. When $a(\cdot,\cdot)$ is not symmetric we can still solve the linear problem [2.10], although it no longer is related to a minimization problem.

Theorem 2. Let $a(\cdot,\cdot)$ be bilinear continuous and V-coercive (see 1.7) on V and let $f \in V'$. Then there is a unique

$$x \in V : a(x,y) = f(y) \ , \quad y \in V \ . \tag{2.11}$$

Proof: For each $x \in V$ the function "$y \longmapsto a(x,y)$" belongs to V', so by Corollary 3 there is a unique $\alpha(x) \in V : (\alpha(x),y) = a(x,y)$, $y \in V$. This defines $\alpha \in \mathcal{L}(V,V)$ and we similarly construct $\beta \in \mathcal{L}(V,V)$ with $(x,\beta(y)) = a(x,y)$ for $x,y \in V$. Since [2.11] is equivalent to $\alpha(x) = \tilde{R}^{-1}(f)$, it suffices to show α is invertible. First, α is one-to-one:

$$c_0\|x\|^2 \leq |a(x,x)| = |(\alpha(x),x)| \leq \|\alpha(x)\|\|x\|$$

so $\alpha(x) = 0$ implies $x = 0$. Also, $c_0\|x\| \leq \|\alpha(x)\|$ for all $x \in V$. Second, the range of α, $Rg(\alpha)$, is closed: If $\lim_{n \to \infty} z_n = z$ and $z_n = \alpha(x_n)$, then $c_0\|x_n - x_m\| \leq \|z_n - z_m\|$ so $\{x_n\}$ is Cauchy, hence, convergent to some $x \in V$. But α is continuous, so $\alpha(x) = z \in Rg(\alpha)$. Finally, since $K \equiv Rg(\alpha)$ is a closed subspace, hence $V = Rg(\alpha) \oplus Rg(\alpha)^{\perp}$, we need only show $Rg(\alpha)^{\perp} = \{0\}$. But if $y \in Rg(\alpha)^{\perp}$ then for every $x \in V$, $0 = (\alpha(x),y) = (x,\beta(y))$, so $\beta(y) = 0$. As above, β is one-to-one, so $y = 0$. Thus $Rg(\alpha) = V$.

Finally, we show that the nonlinear problem [2.9] can be resolved for non-symmetric forms.

Theorem 3. Let $a(\cdot,\cdot)$ be a bilinear, continuous and V-elliptic form on V, K a closed, convex and nonempty subset of V. Then for each $f \in V'$ there exists a unique

$$x \in K : a(x,y-x) \geq f(y-x) \ , \quad u \in K \tag{2.12}$$

and the mapping $f \longmapsto x : V' \longrightarrow K$ is continuous.

Proof: Let x_1 and x_2 be solutions corresponding to f_1 and f_2. Then $a(x_1,x_2-x_1) \geq f_1(x_2-x_1)$, $a(x_2, x_1-x_2) \geq f_2(x_1-x_2)$, and we add these to get $a(x_1-x_2,x_1-x_2) \leq (f_1-f_2)(x_1-x_2)$. This gives $\|x_1-x_2\| \leq (1/c_0)\|f_1-f_2\|_{V'}$, from which follows the uniqueness and continuous dependence.

To prove existence, let $r > 0$ and define $F(x) \in V'$ for each $x \in V$ by

$$F(x)(y) = (x,y) - ra(x,y) + rf(y) \ , \quad y \in V \ .$$

Then note that x is a solution of [2.12] if and only if

$$x \in K : (x,y-x) \geq F(x)(y-x) \ , \quad y \in K \ .$$

But this is equivalent to $x = P_K(\underset{\sim}{R}^{-1}F(x))$; so x is characterized as the fixed point of the function $P_K\underset{\sim}{R}^{-1}F$. Now P_K is a contraction, as follows from a special case of our continuity estimate above, and R is an isometric isomorphism, so it suffices to show F is a strict contraction. But we have

$$|(F(x_1) - F(x_2))(y)| = |(x_1-x_2,y) - r(\alpha(x_1-x_2),y)|$$

where $\alpha : V \to V$ was constructed in Theorem 2, and

$$\|x - r\alpha(x)\|^2 = \|x\|^2 - 2ra(x,x) + r^2\|\alpha(x)\|^2 \leq (1 - 2rc_0 + r^2K^2)\|x\|^2 \ .$$

Choose $r < 2c_0/K^2$ so $\lambda \equiv (1 - 2rc_0 + r^2K^2)^{\frac{1}{2}} < 1$. Then we have $\|F(x_1) - F(x_2)\|_{V'} \leq \lambda\|x_1 - x_2\|$, so it follows that $P_KR^{-1}F$ has a unique fixed point.

III. Function Spaces

We briefly discuss certain aspects of generalized derivatives of functionals and of Sobolev spaces of functions. Our terminology for distributions is non-standard; we refer to any linear functional (not necessarily continuous) on test functions as a distribution. Since all analysis is done in Hilbert subspaces of such functionals, no topological notions are needed for the whole space of functionals.

Let G be a domain in $\mathbb{R}^n$. We say G has the <u>segment property</u> if there is a locally finite open cover $\{G_j\}$ of the boundary ∂G and corresponding nonzero vectors $\{y_j\}$ such that if $x \in \overline{G} \cap G_j$ then $x + ty_j \in G$ for $0 < t < 1$. Also, G has the <u>cone property</u> if there is a cone such that each $x \in G$ is the vertex of some congruent cone contained in G. Finally, G has the <u>uniform C^m-regularity property</u> if there is a locally finite open cover $\{G_j\}$ of ∂G and corresponding C^m diffeomorphisms of G_j onto the unit ball $B \equiv \{\vec{x} \in \mathbb{R}^n : \|\vec{x}\| < 1\}$ mapping $G_j \cap G$ onto $\{\vec{x} \in B : x_n > 0\}$. These properties will be needed for certain technical results later.

For a general domain G we let $\mathcal{D} \equiv C_0^\infty(G)$ denote the linear space of all infinitely differentiable functions $\phi : G \to \mathbb{R}$, each having compact support in G. A linear functional $T : C_0^\infty(G) \to \mathbb{R}$ will be called a <u>distribution</u> on G and the linear space of all distributions is the algebraic dual $\mathcal{D}^*$ of $\mathcal{D}$. Elements of $\mathcal{D}$ are called <u>test functions</u>.

A function $u : G \to \mathbb{R}$ is locally integrable on G if $u \in L^1(K)$ for every compact $K \subset G$; the space of all such (equivalence classes of) functions is denoted by $L^1_{loc}(G)$. If u is (a representative of) an element of $L^1_{loc}(G)$, it determines a distribution u by

$$\tilde{u}(\phi) = \int_G u\phi \, dx , \quad \phi \in \mathcal{D}.$$

Note that $\tilde{u} \in \mathcal{D}^*$ is independent of the representative. Furthermore this construction defines a linear one-to-one map $u \longmapsto \tilde{u}$ of $L^1_{loc}(G)$ into $\mathcal{D}^*$ whereby we hereafter identify functions with functionals. We call $\{\tilde{u} : u \in L^1_{loc}(G)\}$ the <u>regular</u> distributions. Two examples in $\mathbb{R}$ are the Heaviside functional

$$\tilde{H}(\phi) = \int_0^\infty \phi\ , \quad \phi \in C_0^\infty(\mathbb{R})\ ,$$

obtained from the Heaviside function $H(x) = 1$ if $x > 0$ and $H(x) = 0$ for $x < 0$, and the constant functional

$$T(\phi) = \int_{\mathbb{R}} \phi\ , \quad \phi \in C_0^\infty(\mathbb{R})$$

given by $T = \tilde{1}$. An example of a non-regular distribution is the <u>Dirac functional</u> given by

$$\delta(\phi) = \phi(0)\ , \quad \phi \in C_0^\infty(\mathbb{R})\ .$$

For each multi-index of integers, $\alpha = (\alpha_1, \alpha_2, \ldots, \alpha_n)$, we denote the partial derivative of the function $u: G \to \mathbb{R}$ by

$$D^\alpha u(x) = \frac{\partial^{|\alpha|} u(x)}{\partial x_1^{\alpha_1} \ldots \partial x_n^{\alpha_n}}\ , \quad x = (x_1, \ldots, x_n)\ ,$$

where $|\alpha| \equiv \alpha_1 + \alpha_2 + \ldots + \alpha_n$ is the order. We want to extend the derivative to $\mathcal{D}^*$, hence, to $L^1_{loc}(G)$; to be consistent with D^α and with the identification of $L^1_{loc} \subset \mathcal{D}^*$ above, we must have $\partial^\alpha \tilde{u} = \widetilde{(D^\alpha u)}$ for $u \in C^m(G)$ and $|\alpha| \leq m$. That is, by an integration-by-parts it follows

$$\partial^\alpha \tilde{u}(\phi) = (-1)^{|\alpha|} \tilde{u}(D^\alpha \phi)\ , \quad \phi \in \mathcal{D}$$

so we must define the generalized <u>derivative</u> of $T \in \mathcal{D}^*$ by

$$\partial^\alpha T(\phi) = (-1)^{|\alpha|} T(D^\alpha \phi)\ , \quad \phi \in \mathcal{D}\ .$$

Note that D^α is a linear map from $\mathcal{D}$ to itself and ∂^α is $(-1)^\alpha$ times its dual, hence, a linear map from $\mathcal{D}^*$ to itself.

We briefly mention some results in $G = \mathbb{R}$ which are instructive and have immediate extensions to higher dimension.

Lemma. (a) The correspondence $\psi(x) = \int_{-\infty}^{x} \zeta$ establishes the equality of the two sets $\{D\psi : \psi \in \mathcal{D}\}$ and $\{\zeta \in \mathcal{D} : \int \zeta = 0\}$.

(b) Let $\phi_0 \in \mathcal{D}$ with $\int \phi_0 = 1$. Then each $\phi \in \mathcal{D}$ is uniquely written in the form $\phi = \zeta + c\phi_0$ where $\int \zeta = 0$, $\zeta \in \mathcal{D}$, and $c = \int \phi$.

Denote the subspace of $\mathcal{D}$ given in (a) by K. Part (b) says that K is a hyperplane in $\mathcal{D}$.

Theorem 1. (a) For each $S \in \mathcal{D}^*$ there is a (primitive) $T \in \mathcal{D}^*$ with $\partial T = S$.

(b) If $T_1, T_2 \in \mathcal{D}^*$ and $\partial T_1 = \partial T_2$, then $T_1 - T_2$ is a constant in $\mathcal{D}^*$.

(c) Let $T \in \mathcal{D}^*$. Then $\partial T = \tilde{g}$, $g \in L^1_{loc}$, if and only if $T = \tilde{f}$ with f locally absolutely continuous.

Proof: (a) Define T on K by $T(\zeta) = -S(\psi)$, where $\psi(x) = \int_{-\infty}^{x} \zeta \in \mathcal{D}$, and extend to $\mathcal{D}$ by $T(\phi_0) = 0$.

(b) If $\partial T = 0$, then $T(\phi) = T(\zeta + c\phi_0) = T(\phi_0)\tilde{1}$, so T is constant.

(c) If $T = \tilde{f}$, then integration-by-parts shows $\partial T = \widetilde{Df}$. Conversely, if $\partial T = \tilde{g}$, then let f be a primitive of g and note that $\partial(T - \tilde{f}) = 0$. Hence, $T = \tilde{f}$ plus a constant.

Note that part (c) shows that any distribution whose derivative is a function must itself be a function.

Given G in $\mathbb{R}^n$ and a $p \geq 1$, we denote by $L^p(G)$ the class of all (equivalence classes of) measurable functions u on G for which

$$\|u\|_{L^p} \equiv \left(\int_G |u(x)|^p dx\right)^{1/p} < \infty .$$

This defines a norm for which $L^p(G)$ is a Banach space. Each of the generalized derivatives ∂^α is a closed operator in $L^p(G)$. That is, if $u_n \to u$ and $\partial^\alpha u_n \to v$ in $L^p(G)$ then we can let $n \to \infty$ in the identities

$$\partial^\alpha \tilde{u}_n(\phi) = -\tilde{u}_n(D^\alpha \phi)$$

to obtain $\partial^\alpha u = v$ in $L^p(G)$. Hereafter we shall not distinguish $u \in L^1_{loc}$ from $\tilde{u} \in \mathcal{D}^*$ or ∂^α from D^α. We define for each integer $m \geq 0$ and $p \geq 1$ the Sobolev space

$$W^{m,p}(G) \equiv \{u \in L^p(G) : \partial^\alpha u \in L^p(G) \ , \quad |\alpha| \leq m\} \ .$$

Since L^p is complete and each ∂^α is closed, it follows easily that $W^{m,p}$ is a Banach space with the norm

$$\|u\|_{m,p} \equiv \left(\sum_{|\alpha| \leq m} \|\partial^\alpha u\|^p_{L^p} \right)^{1/p} .$$

For the case $p = 2$ we obtain a Hilbert space denoted by $H^m(G) = W^{m,2}(G)$ with the scalar product

$$(u,v)_m = \sum_{|\alpha| \leq m} (\partial^\alpha u, \partial^\alpha v)_{L^2} \ .$$

Since generalized and classical derivatives coincide on $C^m(G)$ it is clear that

$$\hat{C}^{m,p} = \{u \in C^m(G) : \|u\|_{m,p} < \infty\}$$

is a subspace of $W^{m,p}$. Moreover, the following approximation results are known.

Theorem 2. The completion of $\hat{C}^{m,p}$ equals $W^{m,p}$. If G has the segment property, then the set of restrictions to G of functions in $C_0^\infty(\mathbb{R}^n)$ is dense in $W^{m,p}(G)$.

Results of the following kind are very important for a variety of applications, especially to nonlinear problems, and they are known generically as Sobolev imbedding theorems.

Theorem 3. Let G have the cone property in $\mathbb{R}^n$. If $mp < n$ then $W^{m,p}(G) \subset L^q(G)$ for $p \leq q \leq np/(n-mp)$. If $mp = n$ this inclusion holds for $p \leq q < \infty$. If $mp > n$ then $W^{m,p}(G) \subset C_B(G)$, the continuous and bounded functions on G.

We remark that with a Lipschitz-type of smoothness of the boundary ∂G, the last case $mp > n$ can be improved to obtain Hölder-continuity. Furthermore, many applications in analysis, such as the discreteness of the spectra of linear elliptic operators, depend on the compactness of these imbeddings. These are known as Rellich-Kandorachov results.

Theorem 3'. In the situation of Theorem 3, the imbeddings are compact if, in addition, G is bounded and $q < np/(n-mp)$.

An important notion for any discussion of boundary-value problems is the sense in which a function $u: G \to \mathbb{R}$ has a trace or boundary values, $u(s)$, $s \in \partial G$. If $u \in C(\bar{G})$ the meaning is clear, whereas for $u \in L^p(G)$ the values on the null set ∂G are not at all determined. We shall show that the trace, or restriction to the boundary $\gamma_0(u) \equiv u|_{\partial G}$ is defined in $L^p(\partial G)$ whenever $u \in W^{1,p}(G)$.

First we recall the definition of surface integral over ∂G; this leads directly to the definition of $L^p(\partial G)$. Let G have the uniform C^1-regularity property, let $\psi_j : B \longrightarrow G_j$ be the corresponding diffeomorphisms from the unit ball with $\psi_j(B_0) = G_j \cap \partial G$, $B_0 = \{x \in B : x_n = 0\}$. Let $\{\beta_j\}$ be a partition-of-unity subordinate to $\{G_j\}: \beta_j \in C_0^\infty(\mathbb{R}^n)$ has support in $G_j, \beta_j \geq 0$, $\Sigma \beta_j(x) = 1$ for $x \in \partial G$. If g is a function on ∂G, then we define the integral of g over ∂G by

$$\int_{\partial G} g(s)\,ds = \sum_j \int_{\partial G \cap G_j} (\beta_j g)\,ds = \sum_j \int_{B_0} (\beta_j g) \circ \psi_j(y',0) J_j(y')\,dy'$$

where $J_j(y')$ is the Jacobian and dy' denotes Lebesgue measure on B_0 in $\mathbb{R}^{n-1}$. By such a localizing (by $\{\beta_j\}$) and flattening (by $\{\psi_j\}$) we define function spaces on ∂G. Thus, if G has the uniform C^m-regularity property we define $W^{m,p}(\partial G)$ to be those $u \in L^p(\partial G)$ such that $(\beta_j u) \circ \psi_j(y',0)$ belongs to $W^{m,p}(\mathbb{R}^{n-1})$ for all j. This is a Banach space with the norm

$$\|u\|_{m,p} = \left\{ \sum_j \|\beta_j u \circ \psi_j(y',0)\|^p_{W^{m,p}(B_0)} \right\}^{1/p} .$$

Using a different cover, diffeomorphisms, etc., will lead to an equivalent norm, hence, the same space.

The preceding construction shows how statements about function spaces are reduced to the special case of the half-space $\mathbb{R}^n_+ = \{x \in \mathbb{R}^n : x_n > 0\}$.

We show now that the trace can be defined on $W^{1,p}(G)$ with $G = \mathbb{R}^n_+$. Let $\phi\colon \mathbb{R}^n_+ \to \mathbb{R}$ be the restriction of an element of $C_0^\infty(\mathbb{R}^n)$; such functions are dense in $W^{1,p}(\mathbb{R}^n_+)$. Then $\gamma_0(\phi) \equiv \phi|_{\mathbb{R}^n_0}$ with $\mathbb{R}^n_0 = \{x \in \mathbb{R}^n : x_n = 0\}$ defines γ_0 from such ϕ into $C_0^\infty(\mathbb{R}^n_0)$. Since ϕ has compact support we have

$$|\phi(x',0)|^p = -\int_0^\infty D_n|\phi(x',x_n)|^p dx_n = -p\int_0^\infty |\phi(x)|^{p-1}\mathrm{sgn}(\phi)D_n\phi(x)\,dx_n .$$

Integrating over $\mathbb{R}^{n-1}$ leads to

$$\|\gamma_0(\phi)\|^p_{L^p(\mathbb{R}^{n-1})} \le p\int_{\mathbb{R}^n_+} |\phi(x)|\,|D_n\phi(x)|\,dx \le p\|\phi\|^{p/p'}_{L^p(\mathbb{R}^n_+)}\|D_n\phi\|_{L^p(\mathbb{R}^n_+)}$$

$$\le (p-1)\|\phi\|^p_{L^p(\mathbb{R}^n_+)} + \|D_n\phi\|^p_{L^p(\mathbb{R}^n_+)} .$$

Thus γ_0 extends by continuity to $\gamma_0\colon W^{1,p}(\mathbb{R}^n_+) \longrightarrow L^p(\mathbb{R}^{n-1})$. This shows the first trace is well defined and it is clear that its range is dense in $L^p(\mathbb{R}^{n-1})$.

The full trace on $W^{m,p}(G)$ is defined as follows. Let $\phi \in C_0^\infty(\mathbb{R}^n)$; restrictions to G of such functions are dense in $W^{m,p}(G)$. Let γ denote the linear mapping

$$\phi \longmapsto \gamma\phi = (\gamma_0\phi, \gamma_1\phi, \ldots, \gamma_{m-1}\phi) , \qquad \gamma_j = \gamma_0\left(\frac{\partial^j\phi}{\partial\nu^j}\right)$$

where $\frac{\partial}{\partial\nu}$ denotes the directional derivative along the unit outward normal vector ν on ∂G. For $1 < p < \infty$ this extends by continuity to the <u>trace operator</u>

$$\gamma\colon W^{m,p}(G) \longrightarrow \prod_{k=0}^{m-1} W^{m-k-1,p}(\partial G) .$$

The kernel of γ, those $u \in W^{m,p}$ for which $\gamma u = 0$, is characterized as the subspace $W_0^{m,p}(G)$, the closure in $W^{m,p}(G)$ of $C_0^\infty(G)$. For the case $p = 2$, the corresponding Hilbert space is denoted by $H_0^m(G)$.

The preceding results are adequate for a discussion of boundary-value problems with homogeneous boundary conditions. In order to specify the optimal class of functions on ∂G which are allowable boundary values for non-homogeneous problems, it is necessary to characterize the range of γ, $Rg(\gamma)$. This characterization is considerably more difficult than the above, and it requires the spaces of _fractional order_. These spaces are usually constructed directly by "interpolating" between the integer-valued cases $W^{m-1,p}(G)$, $W^{m,p}(G)$. However they can be described directly in terms of a norm involving first-order difference-quotients.

Thus, let $s = m+\sigma$ where $m \geq 0$ is integer and $0 < \sigma < 1$. The norm is given by

$$\|u\|_{s,p} = \left\{ \|u\|^p_{m,p} + \sum_{|\alpha| = m} \int_G \int_G \frac{|\partial^\alpha u(x) - \partial^\alpha u(y)|^p}{|x-y|^{n+\sigma p}} dxdy \right\}^{1/p}$$

and $W^{s,p}(G)$ is the completion of restrictions to G of functions in $C_0^\infty(\mathbb{R}^n)$ with this norm. Since G has the uniform C^m-regularity property, we can define the spaces $W^{s,p}(\partial G)$ as above.

Theorem 4. Let $1 < p < \infty$ and G have the uniform C^m-regularity property. The trace operator γ is a homeomorphism of $W^{m,p}(G)$ onto

$$\prod_{k=0}^{m-1} W^{m-k-1/p,p}(\partial G)$$

and its kernel is $W_0^{m,p}(G)$.

IV. Boundary-Value Problems

We shall construct a variety of examples of elliptic boundary-value problems which can be given in the weak formulation [2.9] or [2.10]. Here we are concerned with the precise interpretations of the weak formulation with specific choices of Sobolev spaces, bilinear forms and linear functionals in the abstract existence results of Part II. Sufficient conditions for these existence results to apply will be presented in Part V along with corresponding approximation results. These examples should include most of the types of problems one would expect to arise from applications.

Let G be a domain in $\mathbb{R}^n$ and suppose we are given a set of coefficient functions which satisfy

$$a_0, a_{ij} \in L^\infty(G) , \quad 1 \leq i, j \leq n \tag{4.1}$$

$$\sum_{i,j=1}^{n} a_{ij}(x)\xi_i\xi_j \geq c(\xi_1^2 + \ldots + \xi_n^2) , \quad \xi \in \mathbb{R}^n, \ x \in G \tag{4.2}$$

$$a_0(x) \geq c , \quad x \in G \tag{4.3}$$

where $c > 0$. Then the bilinear form defined by

$$a(u,v) \equiv \sum_{i,j=1}^{n} \int_G a_{ij}\partial_i u \partial_j v dx + \int_G a_0 uv dx , \quad u,v \in H^1(G) \tag{4.4}$$

is continuous on $H^1(G)$; it is $H^1(G)$-elliptic because of [4.2] and [4.3]. The boundedness of a_0 can be relaxed somewhat by using the Sobolev imbedding theorem; the lower estimates [4.2] and [4.3] will be relaxed below in certain cases depending on our choice of a subspace V of $H^1(G)$. Let $F \in L^2(G)$ be given and define

$$f(v) = \int_G Fv dx , \quad v \in H^1(G) .$$

The form [4.4] is symmetric, so for each choice of a closed subspace V of $H^1(G)$ we obtain from Theorem II.1a the existence of a unique

$$u \in V : a(u,v) = f(v) \ , \quad v \in V \ . \qquad [4.5]$$

In order that [4.5] yield a partial differential equation in $\mathcal{D}^*(G)$, the space V must contain $C_0^\infty(G)$, hence, also its closure, $H_0^1(G)$.

<u>Dirichlet Problem</u>. We choose $V = H_0^1(G)$ and interpret [4.5]. Since $C_0^\infty(G)$ is dense in $H_0^1(G)$, [4.5] is equivalent to

$$u \in H_0^1(G) \qquad [4.6]$$

$$-\sum_{i,j=1}^{n} \partial_j(a_{ij}\partial_i u) + a_0 u = F \quad \text{in } \mathcal{D}^*(G) \qquad [4.7]$$

The partial differential equation [4.7] is <u>elliptic</u> because of [4.2]. The inclusion [4.6] is a generalized Dirichlet-type boundary condition: u vanishes on the boundary ∂G in the sense of trace: $\gamma_0(u) = 0$. Thus [4.6], [4.7] is the homogeneous Dirichlet problem, or boundary-value problem of <u>first</u> <u>type</u>. A corresponding problem with non-homogeneous boundary data is resolved in the form of II[2.9]. Suppose in addition to the above we are given $g \in H^{\frac{1}{2}}(\partial G)$, i.e., $\gamma_0(w) = g$ for some $w \in H^1(G)$. Setting $K \equiv \{w + v : v \in H_0^1(G)\}$, the translate of $H_0^1(G)$ in $H_0^1(G)$ by w, we obtain exactly one

$$u \in K : a(u,v) = f(v) \ , \quad v \in H^1{}_0(G) \ . \qquad [4.8]$$

(Compare II[2.9] and note that $v \in K$ if and only if $v - u \in H_0^1(G)$.) As before, [4.8] is equivalent to [4.7] and

$$u \in H^1(G), \ \gamma_0(u) = g \quad \text{in} \quad H^{\frac{1}{2}}(\partial G) \ . \qquad [4.9]$$

The equation in [4.9] is a non-homogeneous Dirichlet boundary condition.

<u>Neumann Problem</u>. Let's interpret [4.5] with the choice of $V = H^1(G)$. Then we have $u \in H^1(G)$ and (since the equality in [4.5] holds for

$v \in C_0^\infty(G)$) we obtain [4.7]. But $C_0^\infty(G)$ is not dense in $H^1(G)$ so [4.7] is only part of the information in [4.5]. If we substitute [4.7] in [4.5] we obtain after cancellation

$$\sum_{i,j=1}^{n} \int_G a_{ij}\partial_i u \partial_j v dx + \int_G \left\{ \sum_{i,j=1}^{n} \partial_j(a_{ij}\partial_i u) \right\} vdx = 0 , \quad v \in V .$$

Note this is meaningful: [4.7] implies the term in brackets belongs to $L^2(G)$. Assume we know $u \in H^2(G)$. If in addition ∂G is smooth, then the classical Green's formula shows that

$$\int_G \sum_{i,j=1}^{n} \{a_{ij}\partial_i u \partial_j v + \partial_j(a_{ij}\partial_i u)v\} dx \qquad [4.10]$$

$$= \int_{\partial G} \frac{\partial u}{\partial \nu_A} \gamma_0(v) dx , \quad u \in H^2(G),\ v \in H^1(G) ,$$

where $\nu = (\nu_1, \nu_2, \ldots, \nu_n)$ is the unit outward normal and

$$\frac{\partial u}{\partial \nu_A} = \sum_{i,j} a_{ij} \frac{\partial u}{\partial x_i} \nu_j \quad \text{on} \quad \partial G .$$

Thus, if $u \in H^2(G)$ then [4.5] is characterized by [4.7] and $\frac{\partial u}{\partial \nu_A} = 0$ on ∂G. This is the homogeneous Neumann problem, or boundary-value problem of second type.

There remains the proviso above that $u \in H^2(G)$; we were given only $u \in H^1(G)$ by [4.5]. The first way around this is to appeal to the following regularity theorem.

Theorem 1. Assume the domain G is bounded and has the uniform C^{2+k}-regularity property, the coefficients satisfy $a_{ij} \in C^{1+k}(\bar{G})$, and $F \in H^k(G)$ for some integer $k \geq 0$. Then the solution of the Neumann problem belongs to $H^{2+k}(G)$.

Thus the case $k = 0$ in Theorem 1 justifies the use of the classical Green's formula [4.10] to establish the characterization of [4.5] with $V = H^1(G)$ as the Neumann problem. Now Theorem 1 is a typical regularity theorem for elliptic boundary-value problems; a similar result holds for the Dirichlet problem. However there are situations where we do not have smooth boundaries, smooth coefficients or the appropriate type of boundary conditions to obtain such a regularity result. An alternative is to extend the Green's formula to cover the case of all $u \in V$ for which $\sum_{i,j} \partial_j(a_{ij}\partial_i u) \in L^2(G)$. This is guaranteed by [4.7].

As we develop the abstract Green's formula below, it is instructive to compare with the Dirichlet and Neumann problems. The bilinear form [4.4] is equivalent to an operator $\mathcal{A} \in \mathcal{L}(V,V')$ with $V = H^1(G)$ given by

$$\mathcal{A}u(v) = a(u,v) \ , \quad u,v \in V \ .$$

The partial differential equation [4.7] was obtained from the restriction of $\mathcal{A}u$ to $V_0 \equiv H_0^1(G)$. Finally, we note that V_0 is the kernel of the trace operator γ_0 which maps $V \equiv H^1(G)$ onto $H^{\frac{1}{2}}(\partial G)$. What is needed is a "pivot" space common to V' and V_0' so that one may compare $\mathcal{A}u$ with its restriction $\mathcal{A}u|_{V_0}$ and thereby obtain [4.10].

<u>Theorem 2</u>. Let V and B be normed linear spaces and $\gamma \in \mathcal{L}(V,B)$ a strict homomorphism onto B with kernel $V_0 = \ker(\gamma)$. Thus $\gamma^*(g) = g\circ\gamma$ defines an isomorphism of the dual B' onto the annihilator $V_0^\perp \equiv \{f \in V' : f|_{V_0} = 0\}$ in V'. Let X be the space V with a continuous norm $|\cdot|^0$ for which V_0 is dense in X. Thus $X' \subset V'$ and we can identify $X' \subset V_0'$ by restriction. Let $\mathcal{A} \in \mathcal{L}(V,V')$ and define the restriction $A \in \mathcal{L}(V,V_0')$ by $Au = \mathcal{A}u|_{V_0}$, $u \in V$. Define $D = \{u \in V : Au \in X'\}$. Then there is a unique $\partial_A \in \mathcal{L}(D,B')$ for which

$$\mathcal{A}u(v) = Au(v) + \partial_A u(\gamma v) \ , \quad u \in D, \ v \in V \ . \qquad [4.11]$$

Proof: For each $u \in D$ we have $\mathcal{A}u - Au \in V_0^\perp$. Since γ^* is an isomorphism, there is a $\partial_A u \in B'$ for which $\mathcal{A}u - Au = \gamma^*(\partial_A u)$. To check the continuity, note that $u_n \to u$ in D means $u_n \to u$ in V and $Au_n \to Au$ in X'. Hence, $\mathcal{A}u_n \to \mathcal{A}u$ and $Au_n \to Au$ in V', so $(\mathcal{A} - A)u_n \to (\mathcal{A} - A)u$ in $V_0^\perp$. Since γ^* is an isomorphism of B' onto $V_0^\perp$, it follows $\partial_A u_n \to \partial_A u$ in B'.

Before continuing our examples, we note the following. First, the result [4.11] can be obtained without linearity of $\mathcal{A}$; of course A and ∂_A are possibly nonlinear. Second, the construction has no dependence on norms; the spaces could be general topological vector spaces. And finally, the formula is most useful when D, hence, X', are as large as possible. This means the norm $|\cdot|$ (or the topology on X) is the strongest possible permitted in Theorem 2.

For a general example, let the boundary of G be given as the disjoint union $\partial G = \Gamma_0 \cup \Gamma_1$ and define $V \equiv \{u \in H^1(G): \gamma_0(u) = 0$ a.e. on $\Gamma_0\}$. Given $a, a_0 \in L^\infty(G)$ and $a_1 \in L^\infty(\Gamma_1)$ we define the bilinear form

$$a(u,v) = \int_G (a(x)\nabla u(x)\cdot\nabla v(x) + a_0(x)u(x)v(x))\,dx + \int_{\Gamma_1} a_1(s)\gamma_0 u \gamma_0 v\,ds,$$
$$u, v \in V\ . \qquad [4.12]$$

Let X denote V with the $L^2(G)$ norm so $X' = L^2(G)$; also $V_0 = H_0^1(G)$ is the kernel of $\gamma_0: V \to B$, $B \equiv \{\psi \in H^{\frac{1}{2}}(\partial G): \psi = 0$ on $\Gamma_0\}$ is dense and continuously imbedded in $L^2(\Gamma_1)$, so $L^2(\Gamma_1) \subset B'$. The restriction to V_0 of $\mathcal{A}u$ is determined from [4.12] as

$$Au = -\nabla\cdot(a\nabla u) + a_0 u \in \mathcal{D}^*(G)\ . \qquad [4.13]$$

Thus $D = \{u \in V: -\nabla\cdot(a\nabla u) \in L^2(G)\}$ and [4.11] shows that ∂_A is given by

$$\partial_A u(\gamma_0 v) = \int_G (a\nabla u\cdot\nabla v + \nabla\cdot(a\nabla u)v)\,dx + \int_{\Gamma_1} a_1(\gamma_0 u)(\gamma_0 v)\,dx\ ,$$
$$u \in D,\ v \in V\ . \qquad [4.14]$$

If $a \in C^1(\bar{G})$ then $H^2(G) \cap V \subset D$ and for those $u \in H^2(G) \cap V$ we have

$$\partial_A u = a\,\frac{\partial u}{\partial \nu} + a_1\gamma_0(u) \quad \text{in} \quad L^2(\Gamma_1) \subset B'\ . \qquad [4.15]$$

In the "regular" situation of Theorem 1 we have $H^2(G) \cap V = D$, whereas Theorem 2 provides a (possibly proper) extension [4.14] of the boundary values [4.15].

<u>Robin Problem</u>. In addition to the above, suppose we are given $F \in L^2(G)$ and $g \in L^2(\Gamma_1)$; define $f \in V'$ by

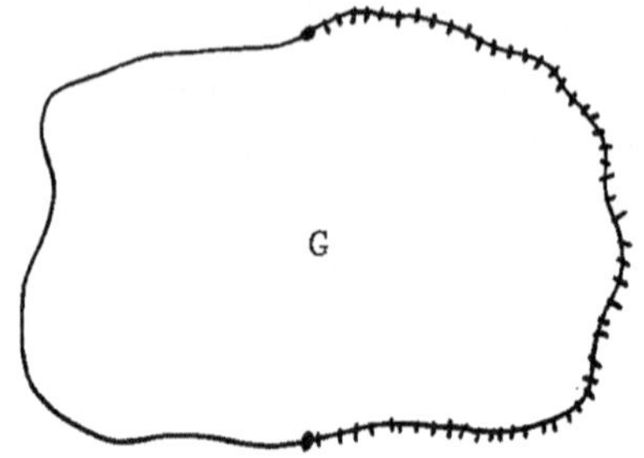

$$f(v) = \int_G F(x)v(x)dx + \int_{\Gamma_1} g(s)\gamma_0 v(s)ds \ , \qquad v \in V \ . \tag{4.16}$$

Let's characterize a solution of [4.5]. First we apply the identity in [4.5] to those $v \in C_0^\infty(G)$ and obtain $Au = F$, hence, $u \in D$. Then from [4.11] we obtain $\partial_A(u) = g$. These calculations are reversible so the variational problem [4.5] is equivalent to

$$\begin{cases} u \in H^1(G),\ Au = F \quad \text{in} \quad L^2(G) \\ \gamma_0 u = 0 \quad \text{in} \quad L^2(\Gamma_0),\ \partial_A u = g \quad \text{in} \quad L^2(\Gamma_1) \end{cases} \tag{4.17}$$

where A and ∂_A are given by [4.13] and [4.14], respectively. The boundary operator [4.15] corresponds to the boundary-value problem of <u>third type</u>, or the Robin problem. It contains the Neumann problem as a special case $(a_1 \equiv 0)$ as well as the Dirichlet problem $(\Gamma_1 = \emptyset)$. The boundary conditions in [4.17] are called <u>mixed</u> when both of Γ_0 and Γ_1 are non-empty.

<u>Adler Problem</u>. We consider [4.12] and [4.16] as above but with $\Gamma_1 = \partial G$, and we let $V \equiv \{u \in H^1(G) : \gamma_0(u)$ is constant, a.e. on $\partial G\}$. (Note that the constant depends on u.) The partial differential operator is given by [4.13] as before. When $a \in C^1(\bar{G})$ we have for those $u \in H^2(G) \cap V$

$$\partial_A u = \int_{\partial G} (a \frac{\partial u}{\partial \nu} + a_1\gamma_0(u))ds \quad \text{in} \quad \mathbb{R} = B' \ .$$

Of course we use [4.14] to evaluate $\partial_A u$ for general $u \in D$. Thus the variational problem [4.5] is equivalent to

$$\begin{cases} u \in H^1(G),\ Au = F \quad \text{in} \quad L^2(G) \\ \gamma_0 u = \text{constant on} \quad \partial G,\ \ \partial_A u = \displaystyle\int_{\partial G} g\,ds \end{cases} \qquad [4.18]$$

with A and ∂_A as above. The pair of boundary conditions in [4.18] are known as <u>fourth-type</u>, and [4.18] is the Adler problem. Such a problem describes the steady-state temperature distribution u of a body G surrounded by a finite reservoir at constant (unknown) temperature $\gamma_0 u$, and then $\int_{\partial G} a \frac{\partial u}{\partial \nu} ds$ is the total flux out of G, $\int_{\partial G} a_1 ds \cdot \gamma_0 u$ is heat lost to outside from the reservoir, and $\int_G g ds$ is the total heat supplied to the reservoir. This is an example of a <u>non-local</u> boundary condition.

<u>Interior Singularity</u>. Suppose the domain G contains a manifold Γ_1 of dimension $n-1$ along which there is a possibility of a singularity in the solution; e.g., let $a(x)$ be smooth at points $x \in G \sim \Gamma_1$ and suppose one-sided limits $a^\pm(x) = \lim\limits_{t \to 0+} a(x \pm t\nu)$ exist for $x \in \Gamma_1$. Here ν denotes the unit normal on Γ_1 for a specified orientation, as well as the outward normal on ∂G. Let [4.12] and [4.16] be given; we shall interpret [4.5] with $V = H^1(G)$. Now the partial differential equation [4.13] will have its classical pointwise meaning in $G \sim \Gamma_1$, so we want

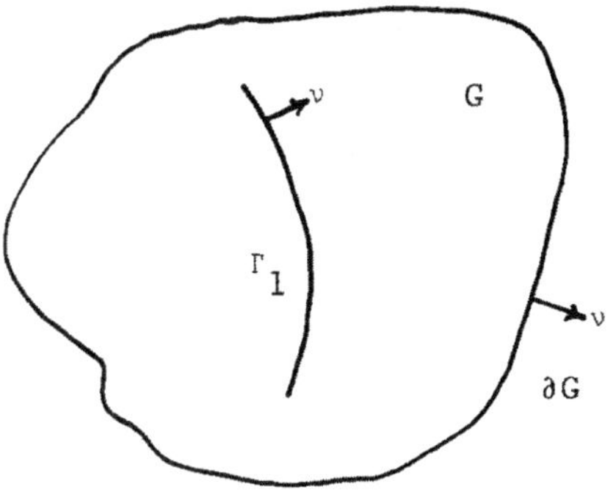

$$V_0 = \{u \in H^1(G) : \gamma_0 u = 0 \quad \text{on} \quad \partial G,\ \gamma_1(u) = C \quad \text{on} \quad \Gamma_1\}$$

where γ_1 denotes the trace on Γ_1, $\gamma_1(u) = u|_{\Gamma_1}$. (This is meaningful since Γ_1 is locally the boundary of that part of G on one side.) Thus we consider the generalized trace

$$\gamma : H^1(G) \longrightarrow H^{\frac{1}{2}}(\partial G) \times H^{\frac{1}{2}}(\Gamma_1) \equiv B$$

defined by $\gamma(u) = [\gamma_0(u), \gamma_1(u)]$ for $u \in H^1(G)$. Setting $v = \phi$ in [4.5] with $\phi \in C_0^\infty(G \sim \Gamma_1)$, we obtain

$$Au = F \quad \text{in } \mathcal{D}^*(G \sim \Gamma_1)$$

where A is given by [4.13]. By using the Green's theorem, from [4.14] we obtain for those $u \in H^2(G \sim \Gamma_1) \cap H^1(G)$

$$\partial_A u = [a \frac{\partial u}{\partial \nu}, (a \frac{\partial u}{\partial \nu})^+ - (a \frac{\partial u}{\partial \nu})^- + a_1\gamma_1(u)] \in L^2(\partial G) \times L^2(\Gamma_1) \subset B' .$$

Thus the solutions of [4.5] are characterized by

$$\begin{cases} u \in H^1(G), \ Au = F \quad \text{in} \quad L^2(G \sim \Gamma_1) \\ a \dfrac{\partial u}{\partial \nu} = 0 \quad \text{on} \quad \partial G \\ (a \dfrac{\partial u}{\partial \nu})^+ - (a \dfrac{\partial u}{\partial \nu})^- + a_1\gamma_1(u) = g \quad \text{on} \quad \Gamma_1 \end{cases} \qquad [4.19]$$

where the last two lines are meant in the sense of $\partial_A u = [0,g]$. Such problems arise in diffusion processes with a discontinuity in the medium (hence, the coefficients) or from a concentrated source (modelled by g) along an interior submanifold. The last equation in [4.19] is known as the <u>transmission condition</u> or <u>interface condition</u>. Note that the identity

$$u^+ = u^- \quad \text{on} \quad \Gamma_1$$

is implicit in $u \in H^1(G)$.

<u>Fracture Surfaces</u>. Problems arise similar to the preceding but wherein the submanifold Γ is the model of a fracture of width $w(s)$ at $s \in \Gamma$. Fractures are regions of extremely high diffusion rates and represent singularities. For simplicity of presentation, we assume Γ is flat: $\Gamma \subset \mathbb{R}^{n-1} \times \{0\}$. Furthermore, assume $\partial\Gamma$ is a C^1 manifold of dimension $n-2$ with unit outward normal Γ_ν. Set

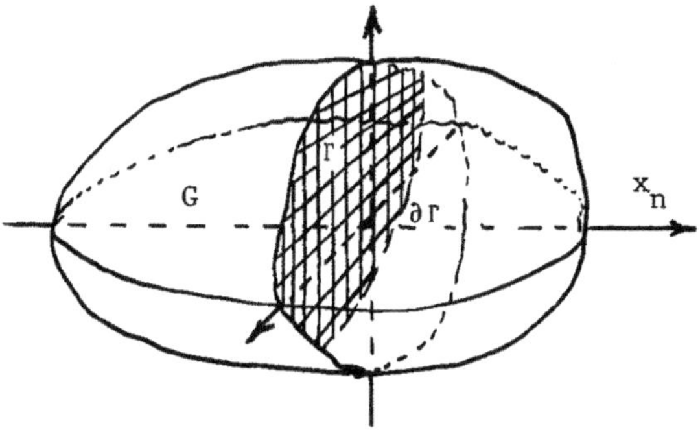

$$V \equiv \{v \in H_0^1(G) : \nabla_0 \gamma_1 v \in L^2(\Gamma)\}$$

where $\gamma_1 v$ is the trace on Γ and ∇_0 is the gradient in the $n-1$ variables $x' = (x_1, x_2, \ldots, x_{n-1})$. For $v \in V$ we have a trace $\gamma_2 v$ on $\partial\Gamma$. Let the given functions $a, a_0 \in L^\infty(G)$ and $w \in L^\infty(\Gamma)$ be positive and define

$$a(u,v) = \int_G (a(x)\nabla u \cdot \nabla v + a_0(x) uv)\, dx + \int_\Gamma w(s) \nabla_0 \gamma_1 u \cdot \nabla_0 \gamma_1 v\, ds \ ,$$

$u, v \in V$.

Let F, F_1 and F_2 be given in L^2 over G, Γ and $\partial\Gamma$, respectively, and

$$f(v) = \int_G Fv\, dx + \int_\Gamma F_1 \gamma_1 v\, ds + \int_{\partial\Gamma} f_2 \gamma_2 v\, dt \ , \qquad v \in V \ .$$

As before we have $V_0 = \{v \in H_0^1(G) : \gamma_1(v) = 0 \text{ on } \Gamma\}$ and a generalized trace

$$\gamma = \gamma_1 : V \longrightarrow H^{\frac{1}{2}}(\Gamma) \equiv B \ .$$

To interpret a solution u of [4.5], we first compute Au in $\mathcal{D}^*(G \sim \Gamma)$ as given by [4.13]. For those $u \in V$ with $u \in H^2(G \sim \Gamma)$ we have

$$\partial_A u(\gamma v) = \int_\Gamma \{((a \frac{\partial u}{\partial \nu})^+ - (a \frac{\partial u}{\partial \nu})^-)v + w(s)\nabla_0 u \cdot \nabla_0 v\} ds , \qquad v \in V .$$

If, in addition, $w \in C^1(\Gamma)$ and $\gamma_1 u \in H^2(\Gamma)$ we have

$$\partial_A u(\gamma v) = \int_\Gamma ((a \frac{\partial u}{\partial \nu})^+ - (a \frac{\partial u}{\partial \nu})^- - \nabla_0 \cdot (w \nabla_0 u))\gamma_1 v ds$$
$$+ \int_{\partial\Gamma} (w \frac{\partial u}{\partial \nu})\gamma_2 v dt , \qquad v \in V .$$

This computation describes the restriction of ∂_A to smooth $u \in D$ as above, and shows that [4.5] is to be interpreted as a weak formulation of the problem

$$\begin{cases} u \in H_0^1(G),\ Au = F \quad \text{in} \quad L^2(G \sim \Gamma) , \\ \nabla_0 u \in H^1(\Gamma),\ (a \frac{\partial u}{\partial \nu})^+ - (a \frac{\partial u}{\partial \nu})^- - \nabla_0(w \nabla_0 u) = F_1 \quad \text{in} \quad L^2(\Gamma) , \\ w \frac{\partial u}{\partial \nu} = F_2 \quad \text{in} \quad L^2(\partial\Gamma) . \end{cases} \qquad [4.20]$$

In the absence of such smoothness of the solution, this is precise in the sense of the abstract Green's formula [4.11].

V. Existence and Approximation

We have just shown that a number of boundary-value problems can be characterized as variational problems in Hilbert space. Now we give sufficient conditions for the abstract existence-uniqueness results of Part II to apply to these examples described in Part IV; one of these conditions is that the partial differential equation be elliptic. Finally we shall recall the general Galerkin method and describe various estimates on the rate of convergence in the more general situations.

Let's begin with the general example following Theorem IV.2. Thus, G is a domain in $\mathbb{R}^n$ whose boundary is a disjoint union $\partial G = \Gamma_0 \cup \Gamma_1$ and $V = \{v \in H^1(G) : \gamma_0(v) = 0$ a.e. on $\Gamma_0\}$. We are given $a, a_0 \in L^\infty(G)$ and $a_1 \in L^\infty(\Gamma_1)$ and define

$$a(u,v) = \int_G (a(x)\nabla u(x)\cdot\nabla v(x) + a_0(x)u(x)v(x))\,dx + \int_{\Gamma_1} a_1(s)\gamma_0 u\gamma_0 v\,ds,$$

$$u,v \in V\ . \qquad [5.1]$$

Also we are given $F \in L^2(G)$, $g \in L^2(\Gamma_1)$ and define

$$f(v) = \int_G F(x)v(x)\,dx + \int_{\Gamma_1} g(s)\gamma_0 v(s)\,ds\ , \qquad v \in V\ . \qquad [5.2]$$

The variational equation [4.5] characterizes the weak solution of the boundary-value problem [4.17], i.e.,

$$\begin{cases} -\nabla\cdot(a\nabla u) + a_0 u = F \quad \text{in} \quad L^2(G)\ , \\ u = 0 \quad \text{in} \quad L^2(\Gamma_0)\ , \\ a\dfrac{\partial u}{\partial \nu} + a_1 u = g \quad \text{in} \quad L^2(\Gamma_1)\ . \end{cases} \qquad [5.3]$$

We seek conditions on data in this problem which imply the form [5.1] is V-coercive. Then Theorem 1a asserts the problem [5.3] is well posed.

We shall always assume $a(x) \geq c > 0$ for a.e. $x \in G$ where $c > 0$. Thus, the quadratic form associated with the principal part of [5.1] is elliptic; see [4.2] and below. It remains to find sufficient conditions on a_0, a_1 and V to imply that [5.1] is V-coercive. It is not enough to have even $a_0 = 0$ and $a_1 = 0$, for in the case $\Gamma_0 = \emptyset$ we may set $v(x) = 1$ in [4.5] to obtain the necessary condition

$$\int_G F(x)\,dx + \int_{\partial G} g(s)\,ds = 0$$

for existence of a solution of the Neumann problem [5.3]. Moreover, non-uniqueness follows by adding a constant to any solution. We give two methods by which one can obtain a coercive estimate, the first by direct calculus and the second by compactness. Both involve either making [5.1] larger (by increasing a_0 or a_1) or making V smaller (by increasing Γ_0).

Theorem 1. Let the domain G be bounded in some direction: there is a $K > 0$ such that $0 \leq x_n \leq K$ for all $x = (x',x_n) \in G$, and suppose ∂G has the uniform C^1-regularity property. Denote the unit outward normal on ∂G by $\nu = (\nu_1, \nu_2, \ldots, \nu_n)$ and define $\Sigma = \{s \in \partial G : \nu_n(s) > 0\}$. Then

$$\int_G |u|^2 dx \leq 2K \int_\Sigma |\gamma_0 u(s)|^2 ds + 4K^2 \int_G |\partial_n u|^2 dx\,, \qquad u \in H^1(G)\,. \tag{5.4}$$

Proof: By Theorem III.2 we may assume u is smooth. Then Gauss' Theorem gives

$$\int_{\partial G} \nu_n(s) s_n |u(s)|^2 ds = \int_G \partial_n (x_n |u(x)|^2)\,dx =$$

$$= \int_G |u|^2 + \int_G x_n \partial_n |u(x)|^2 dx\,.$$

From here it follows

$$\int_G |u|^2 \leq K \int_{\partial G} |u|^2 ds + (\tfrac{1}{2}) \int_G |u|^2 + 2K^2 \int_G |\partial_n u|^2$$

and this gives [5.4].

Corollary.

$$\|u\|_{L^2(G)} \leq 2K\|\partial_n u\|_{L^2(G)} , \quad u \in H_0^1(G) . \tag{5.5}$$

The estimate [5.5] is known as the Poincaré inequality. A consequence is that [5.1] is $H_0^1(G)$-elliptic whenever

$$\operatorname*{ess\,inf}_{x \in G} a_0(x) > -c/2K \tag{5.6}$$

and then the Dirichlet problem ([5.3] with $\Gamma_1 = \emptyset$) is well posed. More generally, from [5.4] it follows that [5.1] is V-elliptic and hence the mixed Dirichlet-Neumann-Robin problem [5.3] is well posed if [5.6],

$$\operatorname*{ess\,inf}_{s \in \Gamma_1} a_1(s) \geq 0$$

and $\Sigma \subset \Gamma_0$. Other combinations are possible and for each we actually calculate a corresponding (non-optimal) coercivity constant, hence, a modulus of continuous dependence of the solution u on the data [5.2]. Similar statements hold for the Adler problem [4.18] when $a_1(s) \equiv a_1 > 0$ for $s \in \partial G = \Gamma_1$. Finally, we note that [5.1] is H^1-elliptic if

$$\begin{aligned} &a(x) \geq c, \ a_0(x) \geq 0, && \text{a.e. } x \in G \\ &a_1(s) \geq 0, && \text{a.e. } s \in \partial G \\ &a_1(s) \geq c, && \text{a.e. } s \in \Sigma \end{aligned}$$

for some constant $c > 0$.

Next we show how coercivity can be obtained from compactness. This applies in the situation of Theorem III.3' where G is bounded and the lower-order terms are compact perturbations.

Theorem 2. Let V be a linear space on which three semi-norms p, q, r are given such that

$$\|x\| \equiv p(x) + r(x) , \qquad |x| \equiv p(x) + q(x) , \qquad x \in V$$

are norms with $\|\cdot\|$ stronger than $|\cdot|$. Assume $\{V,\|\cdot\|\}$ is a reflexive Banach space on which $r(\cdot)$ is compact. Then $\|\cdot\|$ and $|\cdot|$ are equivalent.

Proof: Otherwise, there is a sequence $\{v_n\}$ in V for which $|v_n| \to 0$ and $\|v_n\| = 1$ for $n \geq 1$. Since $\{V,\|\cdot\|\}$ is reflexive and $r(\cdot)$ is compact, there is a subsequence (denoted again by $\{v_n\}$) with weak $\lim_{n\to\infty} v_n = v$ and $\lim_{n\to\infty} r(v_n) = r(v)$. But then weak $\lim v_n = v$ in $\{V,|\cdot|\}$ so $v = 0$ by uniqueness of weak limits, and $p(v_n) \to 0$, $r(v_n) \to 0$ contradicting $\|v_n\| = 1$.

<u>Corollary</u>. Let G be a bounded domain in $\mathbb{R}^n$ which has the cone property. Let the bilinear form [5.1] be given with (essentially) bounded coefficients which also satisfy $a(x) \geq c > 0$, $a_0(x) \geq 0$ for a.e. $x \in G$ and $a_1(s) \geq 0$ for a.e. $s \in \Gamma_1$. Then a necessary and sufficient condition for [5.1] to be H^1-elliptic is that $a_0(x) > 0$ on a set of strictly positive measure or that $a_1(s) > 0$ on a set of strictly positive measure.

Proof: The necessity follows from our preceding remarks on the Neumann problem. The sufficiency follows from Theorem 2 with

$$p(v) = \left(\int_G a(x)|\nabla v|^2 dx\right)^{\frac{1}{2}}, \quad r(v) = \|v\|_{L^2(G)},$$

$$q(v) = \left(\int_G a_0(x)u^2 dx + \int_{\Gamma_1} a_1(s)|\gamma_0 u|^2 ds\right)^{\frac{1}{2}}$$

and the compactness of the imbedding $H^1(G) \to L^2(G)$.

Similarly one can obtain ellipticity of [5.1] on subspaces of $H^1(G)$: it is then necessary and sufficient to show that $p+q$ is a norm on that subspace, i.e., zero is the only constant function in that subspace for which q vanishes.

Briefly we consider bilinear forms more general than [5.1]. If we are given a collection of functions $a_{ij}(1 \leq i,\ j \leq n)$, $a_j(0 \leq j \leq n)$ in $L^\infty(G)$ we define

$$a(u,v) = \int_G \{\sum_{i,j=1}^{n} a_{ij}\partial_i u \partial_j v + \sum_{j=1}^{n} a_j \partial_j u v + a_o uv\} dx, \quad u,v \in H^1(G). \qquad [5.7]$$

This yields a general second-order linear partial differential operator in divergence form,

$$Au = -\sum_{i,j=1}^{n} \partial_j(a_{ij}\partial_i u) + \sum_{j=1}^{n} a_j\partial_j u + a_0 u . \qquad [5.8]$$

The quadratic form $\{a_{ij}\}$ which determines the principal part of [5.8] is called strongly elliptic at $x \in G$ if for some $c(x) > 0$ we have

$$\sum_{i,j=1}^{n} a_{ij}(x)\xi_i\xi_j \geq c(x) \sum_{j=1}^{n} |\xi_j|^2 , \qquad \xi \in \mathbb{R}^n , \qquad [5.9]$$

and it is uniformly strongly elliptic if [5.9] holds with $c(x) = c > 0$ independent of $x \in G$. Note that these notions are invariant under a change of coordinates. We can duplicate all of our results for [5.1] when the principal part is a uniformly strongly elliptic quadratic form. Moreover we can dominate the first-order terms in [5.7] by adding a large enough multiple of the identity to [5.8]. This is the following very special case of Garding's inequality.

Theorem 3. If [5.7] is uniformly strongly elliptic then there is a $\lambda_0 \in \mathbb{R}$ such that for every $\lambda > \lambda_0$ the bilinear form $a(u,v) + \lambda(u,v)_{L^2(G)}$ is H^1-elliptic.

For our final topic here we present various estimates of the error that results when the Galerkin method is used to approximate the solution of a variational problem with finite-dimensional subspaces having approximation properties typical of finite-element subspaces. A special case was given in Part I; here we show those results are typical for a large class of boundary-value problems in variational form.

Theorem 4. Let $a(\cdot,\cdot)$ be a V-coercive continuous form, i.e., there are constants $K, c > 0$ such that

$$|a(u,v)| \leq K\|u\|\|v\| , \qquad |a(v,v)| \geq c\|v\|^2 , \qquad u,v \in V$$

where $\|\cdot\|$ denotes the norm on Hilbert space V. Let S be a closed subspace of V and $f \in V'$. Then there is exactly one

$$u \in V : a(u,v) = f(v) , \qquad v \in V \qquad [5.10]$$

and exactly one

$$u_S \in S : a(u_S, v) = f(v) , \quad v \in S \tag{5.11}$$

and these satisfy the estimate

$$\|u - u_S\| \leq (K/c) \inf\{\|u - w\| : w \in S\} . \tag{5.12}$$

Proof: The existence and uniqueness are immediate from Theorem II.2. For any $w \in S$ we have

$$a(u_S - u, u_S - u) = a(u_S - u, w - u) + a(u_S - u, u_S - w)$$

and the last term is zero since $v = u_S - w \in S$ can be used in [5.10] and [5.11]. Thus

$$c\|u_S - u\|^2 \leq K\|u_S - u\| \|w - u\|$$

so [5.12] follows.

When V is a subspace of $H^1(G)$, [5.12] is the <u>energy estimate</u> of error. To obtain improved rates of "L^2 estimates" and describe appropriate regularity hypotheses, we introduce a second "pivot" Hilbert space (cf. Theorem IV.2) H such that V is dense and continuously imbedded in H, H is identified with H', so $V \subset H \subset V'$. Thus each $F \in H$ determines $f \in V'$ by $f(v) = (F,v)_H$, $v \in V$; we let D be the set of all corresponding solutions u of [5.10] and denote this by $Au = F$. One should check that this is equivalent to defining $A: D \to H$ by

$$a(u,v) = (Au,v)_H , \quad u \in D, \ v \in V$$

on a maximal domain $D \subset V$. Furthermore, in the situation of Theorem IV.2 with $X = H$ we have $D = \{u \in V : Au \in H \text{ and } \partial_A u = 0\}$, so the domain D is characterized by the complete set of homogeneous boundary conditions for the boundary-value problem. In the situation of Theorem IV.1, we have $D \subset H^2(G)$. Similar remarks hold for the adjoint operator $A^*: D^* \to H$ obtained from the adjoint form,

$$a(u,w) = (u, A^* w) \ , \quad u \in V, \ w \in D^* \ .$$

<u>Theorem 5</u>. Given the above, suppose there is a constant $e^*(S) > 0$ such that

$$\inf\{\|w - v\| : v \in S\} \leq e^*(S)\, |A^* w| \ , \quad w \in D^* \qquad [5.13]$$

where $|\cdot|$ is the norm on H. Then the solutions of [5.10] and [5.11] satisfy

$$|u - u_S| \leq (K^2/c)\ \inf\{\|u - v\| : v \in S\} e^*(S) \ . \qquad [5.14]$$

In addition, suppose there is $e(S) > 0$ such that

$$\inf\{\|w - v\| : v \in S\} \leq e(S)\, |Aw| \ , \quad w \in D \ . \qquad [5.15]$$

Then we have

$$|u - u_S| \leq (K^2/c) e(S) e^*(S) |Au| \ , \quad u \in D \ . \qquad [5.16]$$

Proof: Let $u \neq u_S$ and $g = (u - u_S)/|u - u_S|$. Choose $w \in D^*$ so $Aw = g$. Then

$$a(v,w) = (v,g)_H \ , \quad v \in V$$

and this implies

$$|u - u_S|_H = (u - u_S, g)_H = a(u - u_S, w) = a(u - u_S, w - v) \ , \quad v \in S \ ,$$

hence,

$$|u - u_S| \leq K\|u - u_S\| e^*(S)\, |A^*w| \ .$$

Since $|A^*w| = 1$ the estimate [5.14] follows from [5.12]. Finally, [5.16] follows from [5.14] and [5.15].

We illustrate these estimates for typical finite-element subspaces indexed by a parameter $0 < h \leq 1$ related to the mesh size of a partition of G. Thus, let $\mathcal{S} \equiv \{S_h : 0 < h \leq 1\}$ be closed subspaces of $V \subset H^1(G)$ and $M, k \geq 0$ be integers for which the following <u>approximation assumption</u> holds:

$$\inf\{\|w - v\|_{H^1(G)} : v \in S_h\} \leq M h^{j-1} \|w\|_{H^j(G)}, \quad 0 < h \leq 1,\ 1 \leq j \leq k+2,$$

$$w \in H^j(G) \cap V \ .$$

The integer $k+1$ is the <u>degree</u> of $\mathcal{S}$. Let $f \in V'$, u be the solution of [5.10] and u_h the solution of [5.11] with $S = S_h \in \mathcal{S}$. Suppose the form $a(\cdot,\cdot)$ is "regular" in the sense that Theorem IV.1 holds (with the same $k \geq 0$). Then $|A^*w|_{L^2} \cong \|w\|_{H^2(G)}$, so [5.13] holds with $e^*(S_h) = \mathcal{O}(h)$. From [5.14] we obtain

$$\|u - u_h\|_{L^2} = \mathcal{O}(h) \ , \quad 0 < h \leq 1 \ . \tag{5.17}$$

Suppose further that $f \in H^k(G)$. Then [5.12] implies

$$\|u - u_h\|_{H^1} = \mathcal{O}(h^{k+1}) \ , \quad 0 < h \leq 1 \tag{5.18}$$

since $u \in H^{k+2}(g)$ by Theorem IV.1, and from [5.16] we obtain

$$\|u - u_h\|_{L^2} = \mathcal{O}(h^{k+2}) \ , \quad 0 < h \leq 1 \ . \tag{5.19}$$

Note that we obtained the L^2 estimates [5.14] and [5.17] for general $f \in V'$ and for the smoother case of $f \in H^k(G)$, $k \geq 0$, we obtained the higher convergence rates [5.16], [5.18] and [5.19]. The convergence rate is limited by the smoothness of the "data" in the problem (via Theorem IV.1) and by the degree of the family $\mathcal{S}$ of approximating subspaces of V.

Vol. 1090: Differential Geometry of Submanifolds. Proceedings, 1984. Edited by K. Kenmotsu. VI, 132 pages. 1984.

Vol. 1091: Multifunctions and Integrands. Proceedings, 1983. Edited by G. Salinetti. V, 234 pages. 1984.

Vol. 1092: Complete Intersections. Seminar, 1983. Edited by S. Greco and R. Strano. VII, 299 pages. 1984.

Vol. 1093: A. Prestel, Lectures on Formally Real Fields. XI, 125 pages. 1984.

Vol. 1094: Analyse Complexe. Proceedings, 1983. Edité par E. Amar, R. Gay et Nguyen Thanh Van. IX, 184 pages. 1984.

Vol. 1095: Stochastic Analysis and Applications. Proceedings, 1983. Edited by A. Truman and D. Williams. V, 199 pages. 1984.

Vol. 1096: Théorie du Potentiel. Proceedings, 1983. Edité par G. Mokobodzki et D. Pinchon. IX, 601 pages. 1984.

Vol. 1097: R.M. Dudley, H. Kunita, F. Ledrappier, École d'Éte de Probabilités de Saint-Flour XII – 1982. Edité par P.L. Hennequin. X, 396 pages. 1984.

Vol. 1098: Groups – Korea 1983. Proceedings. Edited by A.C. Kim and B.H. Neumann. VII, 183 pages. 1984.

Vol. 1099: C.M. Ringel, Tame Algebras and Integral Quadratic Forms. XIII, 376 pages. 1984.

Vol. 1100: V. Ivrii, Precise Spectral Asymptotics for Elliptic Operators Acting in Fiberings over Manifolds with Boundary. V, 237 pages. 1984.

Vol. 1101: V. Cossart, J. Giraud, U. Orbanz, Resolution of Surface Singularities. VII, 132 pages. 1984.

Vol. 1102: A. Verona, Stratified Mappings – Structure and Triangulability. IX, 160 pages. 1984.

Vol. 1103: Models and Sets. Proceedings, Logic Colloquium, 1983, Part I. Edited by G.H. Müller and M.M. Richter. VIII, 484 pages. 1984.

Vol. 1104: Computation and Proof Theory. Proceedings, Logic Colloquium, 1983, Part II. Edited by M.M. Richter, E. Börger, W. Oberschelp, B. Schinzel and W. Thomas. VIII, 475 pages. 1984.

Vol. 1105: Rational Approximation and Interpolation. Proceedings, 1983. Edited by P.R. Graves-Morris, E.B. Saff and R.S. Varga. XII, 528 pages. 1984.

Vol. 1106: C.T. Chong, Techniques of Admissible Recursion Theory. IX, 214 pages. 1984.

Vol. 1107: Nonlinear Analysis and Optimization. Proceedings, 1982. Edited by C. Vinti. V, 224 pages. 1984.

Vol. 1108: Global Analysis – Studies and Applications I. Edited by Yu.G. Borisovich and Yu.E. Gliklikh. V, 301 pages. 1984.

Vol. 1109: Stochastic Aspects of Classical and Quantum Systems. Proceedings, 1983. Edited by S. Albeverio, P. Combe and M. Sirugue-Collin. IX, 227 pages. 1985.

Vol. 1110: R. Jajte, Strong Limit Theorems in Non-Commutative Probability. VI, 152 pages. 1985.

Vol. 1111: Arbeitstagung Bonn 1984. Proceedings. Edited by F. Hirzebruch, J. Schwermer and S. Suter. V, 481 pages. 1985.

Vol. 1112: Products of Conjugacy Classes in Groups. Edited by Z. Arad and M. Herzog. V, 244 pages. 1985.

Vol. 1113: P. Antosik, C. Swartz, Matrix Methods in Analysis. IV, 114 pages. 1985.

Vol. 1114: Zahlentheoretische Analysis. Seminar. Herausgegeben von E. Hlawka. V, 157 Seiten. 1985.

Vol. 1115: J. Moulin Ollagnier, Ergodic Theory and Statistical Mechanics. VI, 147 pages. 1985.

Vol. 1116: S. Stolz, Hochzusammenhängende Mannigfaltigkeiten und ihre Ränder. XXIII, 134 Seiten. 1985.

Vol. 1117: D.J. Aldous, J.A. Ibragimov, J. Jacod, Ecole d'Été de Probabilités de Saint-Flour XIII – 1983. Édité par P.L. Hennequin. IX, 409 pages. 1985.

Vol. 1118: Grossissements de filtrations: exemples et applications. Seminaire, 1982/83. Edité par Th. Jeulin et M. Yor. V, 315 pages. 1985.

Vol. 1119: Recent Mathematical Methods in Dynamic Programming. Proceedings, 1984. Edited by I. Capuzzo Dolcetta, W.H. Fleming and T. Zolezzi. VI, 202 pages. 1985.

Vol. 1120: K. Jarosz, Perturbations of Banach Algebras. V, 118 pages. 1985.

Vol. 1121: Singularities and Constructive Methods for Their Treatment. Proceedings, 1983. Edited by P. Grisvard, W. Wendland and J.R. Whiteman. IX, 346 pages. 1985.

Vol. 1122: Number Theory. Proceedings, 1984. Edited by K. Alladi. VII, 217 pages. 1985.

Vol. 1123: Séminaire de Probabilités XIX 1983/84. Proceedings. Edité par J. Azéma et M. Yor. IV, 504 pages. 1985.

Vol. 1124: Algebraic Geometry, Sitges 1983. Proceedings. Edited by E. Casas-Alvero, G.E. Welters and S. Xambó-Descamps. XI, 416 pages. 1985.

Vol. 1125: Dynamical Systems and Bifurcations. Proceedings, 1984. Edited by B.L.J. Braaksma, H.W. Broer and F. Takens. V, 129 pages. 1985.

Vol. 1126: Algebraic and Geometric Topology. Proceedings, Rutgers 1983. Edited by A. Ranicki, N. Levitt and F. Quinn. V, 523 pages 1985.

Vol. 1127: Numerical Methods in Fluid Dynamics. Edited by F. Brezzi. VII, 333 pages. 1985.

Vol. 1129: Numerical Analysis, Lancaster 1984. Proceedings. Edited by P.R. Turner. XIV, 179 pages. 1985.